101
Investment Tools
FOR
Buying Low and Selling High

101
Investment
Tools
FOR
Buying Low
and Selling High

Jae K. Shim, Ph.D.

Jonathan Lansner

S^t_L

St. Lucie Press

Boca Raton   London   New York   Washington, D.C.

Library of Congress Cataloging-in-Publication Data

Shim, Jae K.
    101 investment tools for buying low and selling high / Jae K. Shim, Jonathan Lansner.
        p. cm.
    ISBN 0-91044-13-X
    1. Speculation 2. Stocks 3. Investments I. Title: One hundred one investment tools for buying low and selling high. II. Lansner, Jonathan. III. Title.

    HG6041 .S485 2000
    332.63′2042—dc21   00-46890

© 2001 by CRC Press LLC
St. Lucie Press is an imprint of CRC Press LLC

No claim to original U.S. Government works
International Standard Book Number 0-910944-13-X
Library of Congress Card Number 00-046890
Printed in the United States of America 1 2 3 4 5 6 7 8 9 0
Printed on acid-free paper

The Authors

Jae K. Shim, Ph.D., is Professor of Finance and Accounting at California State University, Long Beach, and President of the National Business Review Foundation, an investment consulting and training firm. He is also Chief Investment Officer (CIO) of a Los Angeles-based investment firm. Dr. Shim received his MBA and Ph.D. degrees from the University of California at Berkeley. He has published numerous articles in professional and academic journals.

Dr. Shim has over 50 business books to his credit, including *Personal Finance, Encyclopedic Dictionary of Accounting and Finance, The Source: Investment Information, The Vest-Pocket Investor, Financial Management,* and the best-selling *Vest-Pocket MBA.* He has been frequently quoted by such media as *The Los Angeles Times, Personal Finance,* and *Money Radio.*

Dr. Shim was the 1982 recipient of the *Credit Research Foundation Award* for his article on investment management.

Jonathan Lansner is the business columnist for *The Orange County Register* in Santa Ana, CA, and editor of the Society of American Business Editors' and Writers' Newsletter, *The Business Journalist.*

A financial journalist since 1983, Lansner has drawn a diverse collection of assignments—everything from award-winning investigative stories on the nation's savings and loan debacle to how to stories on investing basics for novices to covering Orange County's historic municipal bankruptcy.

Before becoming *The Register*'s business columnist in April, he was part of *The Register*'s Business Section's management team for five years. As a business editor, he oversaw *The Register*'s coverage of personal finance, investment markets, the economy and real estate. *The Register*'s coverage of the *Comparator* penny stock scandal, edited by Lansner, won a 1997 Gerald Loeb Award, financial journalism's highest honor.

A 1979 graduate of the University of Pennsylvania's Wharton School of Business, Lansner worked at *The Pittsburgh Press* from 1979 to 1986 and has since been with *The Register.* His last book, *How Money Works* was published in 1995 by Ziff-Davis Press.

Preface

Decades ago, investors scrambled to get the smallest bit of "inside" knowledge from stock, bond, and other investment markets. Stock quotations had to be gleaned from a broker and yields were best found at the bank. A need for detailed statistics required either expensive market newsletters or a trip to the library—or both.

Today, the challenge is very different. The chore is to sort through the vast array of facts. There are figures available in the local newspaper, no less, the national financial newspapers—*The Wall Street Journal, Barron's, Investor's Business Daily*—and in investment magazines—*Forbes, Money, Smart Money, Worth*—on television—*CNNfn, CNBC*—on a home computer and on the Internet. Then, of course, there is the information contained in the many market advisory services that still thrive.

Consider this: A typical major city newspaper contains quotations on more than 8000 stocks, bonds, and mutual funds and more than 100 indexes. Add to that expanded coverage of economic statistics or the added availability of such information on television, radio, and via online information services accessed by home computers. Most investors are not able to decipher what is up from what is down.

101 Investments Tools for Buying Low and Selling High is designed to help the investor cut through all the statistical noise. It gets a reader quickly to the best financial barometers and helps him or her make informed investment decisions.

More than just an ordinary investment dictionary, *101 Investment Tools* analyzes, in a concise style, the various investment vanes—from stock indexes to measures of housing affordability to leading economic indicators. The reader will learn what these measures are, who is compiling them, where they are easily found (publications, media, more) and how they can, or cannot, be used to guide investment decisions. All in a handy, carry-along format.

In the ongoing complex investment climate we live in, those who understand and can use such investment tools will be the ones who succeed.

The authors are grateful to Allison Shim for her enormous word-processing and spreadsheet assistance and Marianne, Rachel, and Jake Lansner for being so understanding during this project.

Jae K. Shim, Los Alamitos, CA
Jonathan Lansner, Trabuco Canyon, CA

Contents

Tool #1 AMEX Major Market Index 1
    Figure 1 AMEX Major Market Index 1

Tool #2 AMEX Composite Index 3

Tool #3 AMEX: Other Indexes 4
    Institutional Index 4
    Eurotop 100 Index. 5
    Inter@Active Week Internet Index 6
        Figure 2 AMEX Inter@Active Week Internet Index 6
    Oil Index 7

Tool #4 Arbitrage 8

Tool #5 ARMS Index (TRIN) 10
    Figure 3 Computing an Arms Index. 10
    Figure 4 The Arms Index 11

Tool #6 Asset Allocation 12
    Figure 5 Asset Allocation Grids. 13
    Figure 6 Portfolio Suggestions by Asset Allocation 14

Tool #7 Barron's Indexes 17
    Confidence Index 17
    50-Stock Average 18
    Figure 7 Barron's 50-Stock Index 19

Tool #8 Beta for a Mutual Fund 20
    Figure 8 Betas for Large Mutual Funds. 21

Tool #9 Beta for a Security 23
    Figure 9 Selected Betas for Stocks 23

Tool #10 Bond Market Indexes 25
    Bond Aggregate Indexes. 25
    Figure 10 Selected Bond Market Indexes 26
    Government Bond Indexes 26
    Corporate Bond Indexes 27

Municipal Bond Indexes 28
    Figure 11 Bond Indexes 29
Global Bond Indexes 29
    Figure 12 Components of Dow Jones 20 Bond Average 31

Tool #11 Breadth (Advance–Decline) Index 33
    Figure 13 Advance–Decline for Stocks 34

Tool #12 British Pound 36
    Figure 14 British Pound 37
    Figure 15 Currency Impact on Investing 38

Tool #13 Bullishness Indicators 39
    Figure 16 How the Bridge Barometer Fluctuates 40

Tool #14 Cash Investments 41
Certificate of Deposit Yields 41
    Figure 17 CD Rates 42
Guaranteed Investment Contract Yields 42
    Figure 18 Guaranteed Investment Contracts 43
Money Market Fund Average Maturity 44
Money Market Fund Yields 44
    Figure 19 Money Market Mutual Fund Results 45

Tool #15 Changes: Stocks Up and Down 47
    Figure 20 Ups and Downs 48

Tool #16 Charting 49
    Figure 21 Line Chart 49
    Figure 22 Bar Chart 50
    Figure 23 Point-and-Figure Chart 50

Tool #17 Commodities Indexes 52
CRB Indexes 52
    Figure 24 CRB Index 53
The Economist Commodities Index 54

Tool #18 Contrarian Investing 55
Contrary Opinion Rule 55
Index of Bearish Sentiment 55
    Figure 25 Contrarian Indicators 57

Tool #19 Credit Ratings 58
    Figure 26 S&P and Moody's Rating Systems 59

Tool #20 Crude Oil Spot Price 60
    Figure 27 Crude Oil Futures Price Chart 60

Tool #21 Currency Indexes 62
    Federal Reserve Trade-Weighted Dollar 62
    J.P. Morgan Dollar Index 62
    Figure 28 Currency 63

Tool #22 Dollar-Cost Averaging 64
    Figure 29 How Dollar-Cost Averaging Works 65

Tool #23 Dow Jones Industrial Average 66
    Figure 30 Dow Jones Industrial Average 67
    Figure 31 Dow Jones Industrial Average Milestones 68

Tool #24 Dow Jones Industry Groups 69
    Figure 32 Dow Jones Industry Groups 70

Tool #25 Dow Jones Global Stock Indexes 71
    Figure 33 Dow Jones Global Indexes 72

Tool #26 Dow Jones Transportation Average 73
    Figure 34 Dow Jones Transportation Index 74

Tool #27 Dow Jones Utilities Average 75
    Figure 35 Dow Jones Utilities Index 76

Tool #28 Duration 77
    Figure 36 Duration: One Calculation 78
    Figure 37 Example of Mutual Fund Duration 79

Tool #29 Economic Indicators and Bond Yields 80
    Figure 38 Probable Effects of Economic Variables on Bond Yields 81

Tool #30 Economic Indicators and Stocks 82
    Figure 39 Economic Variables and Their Impacts
    on the Economy and Stocks 83

**Tool #31 Economic Indicators: Factory Orders and Purchasing
    Manager's Index** 84
    Figure 40 Factory Orders Report 85

Tool #32 Economic Indicators: Gross Domestic Product 86
    Figure 41 Gross Domestic Product 86

**Tool #33 Economic Indicators: Housing Starts
and Construction Spending**. 89
    Figure 42 Housing Starts 90

Tool #34 Economic Indicator: Index of Leading Indicators 91

**Tool #35 Economic Indicators: Industrial Production
and Capacity Utilization** 93
    Figure 43 Industrial Production Report 94

Tool #36 Economic Indicators: Inflation. 95
    Figure 44 Consumer Price Index 96
    Figure 45 Producer Price Index 96
    Figure 46 Employment Cost Index 97

Tool #37 Economic Indicators: Money Supply. 99
    Figure 47 Money Supply Report 99

**Tool #38 Economic Indicators: Personal Income
and Confidence Indices** 101
    Figure 48 Personal Income 102
    Figure 49 Consumer Confidence 102

Tool #39 Economic Indicators: Productivity. 104

Tool #40 Economic Indicator: Recession. 105

Tool #41 Economic Indicator: Retail Sales 106

**Tool #42 Economic Indicators: Unemployment Rate, Initial
Jobless Claims, and Help-Wanted Index**. 107

**Tool #43 Economic Indicators: U.S. Balance of Payments
and the Value of the Dollar** 109
    Figure 50 Foreign Exchange Report 110

Tool #44 The Euro 112

Tool #45 Footnotes on Newspaper Financial Tables. 114

Tool #46 FT-SE "Footsie" 100 (U.K.) Stock Index 117
    Figure 51 FT-SE 100 118

Tool #47 German Mark 119

Tool #48 German Share Index (Frankfurt Dax). 120
  Figure 52 Foreign Market Indicators 121

Tool #49 Gold Spot Price 122
  Figure 53 Precious Metal Prices 123

Tool #50 Gross Income Multiplier (GIM). 124

Tool #51 Herzfeld Closed-End Average. 125
  Figure 54 Herzfeld Closed-End Average 126

Tool #52 Index Arbitrage 127

Tool #53 Indexing 129

Tool #54 Initial Public Offerings (IPOs) 131
  Figure 55 Initial Public Offerings Listings 132

Tool #55 Insider Trading Activity 133
  Figure 56 Insider Activity 134

Tool #56 Interest Rates: Fed Funds, Discount, and Prime 135
  Figure 57 Money Rates Report 136

Tool #57 Interest Rates: 30-Year Treasury Bonds 137

Tool #58 Interest Rates: Three-Month Treasury Bills 138
  Figure 58 T-Bill Auction 139

Tool #59 Investing Styles. 140
  Growth Investing. 140
  Value Investing 141

Tool #60 IBD'S *Smart Select*™ **Ratings**. 143
  Figure 59 Sample of *Smart Select*™ 144

Tool #61 Japanese Candlestick Charts 145
  Figure 60 Candlestick Chart 146

Tool #62 Japanese Yen. 147

Tool #63 Jensen's Performance Measure (Alpha) 148

Tool #64 Lipper Mutual Fund Indexes 151
    Figure 61 Lipper Mututal Fund Indexes 152

Tool #65 Lipper Mutual Fund Rankings 153
    Figure 62 Lipper Mutual Fund Rankings 154

Tool #66 Misery Index 155

Tool #67 Momentum Gauges 156
  Oscillators 156
  Relative Strength Analysis 157
    Figure 63 Relative Strength for Stocks 158
  Summation Index 159
  "Wall Street Week" Technical Market Index 159

Tool #68 Morgan Stanley EAFE Index 161
    Figure 64 Other MSCI Foreign Stock Indexes 162

Tool #69 Morningstar Mutual Fund Rankings 163
    Figure 65 Morningstar Style Boxes 165
    Figure 66 Morningstar Reports 165

Tool #70 Mortgage Rates 167
  Weekly Average 167
  Adjustable Loan Benchmarks 167
    Figure 67 Mortgage Rates Report 168
    Figure 68 Mortgage Benchmark Indexes 169

Tool #71 Most Active Issues 170
    Figure 69 Most Active Stocks 170

Tool #72 Moving Averages 172

Tool #73 Mutual Fund Cash-to-Assets Ratio 174

Tool #74 Mutual Funds: Evaluation Tools 176
  Alpha for a Mutual Fund 176
  Expense Ratios 177
  Mutual Fund Net Asset Value 177
    Figure 70 Mutual Fund Expense Ratios 178
  R^2 for a Mutual Fund 179
  Standard Deviation for a Mutual Fund 180
    Figure 71 Mutual Fund Analytical Tools 182

Tool #75 Mutual Funds: Sales Figures 183

Tool #76 NASDAQ Indexes 184
    Figure 72 NASDAQ Indexes 185

Tool #77 New Highs-to-Lows Ratio 187
    Figure 73 New Highs, New Lows 188

Tool #78 New York Stock Exchange Indexes 189
    Figure 74 New York Stock Exchange Index 190

Tool #79 Nikkei (Tokyo) Stock Index 191
    Figure 75 Nikkei Stock Index 191
    Figure 76 Other Foreign Stock Indexes 193

Tool #80 Option Tools 194
    CBOE Put–Call Ratio.. 194
    Put–Call Options Premium Ratio.............................. 194
    Spread Strategy... 195
    Straddling Strategy 196
    Valuation of Options (Calls and Puts)......................... 196
    Valuation of Stock Warrants 201
    Value of Stock Rights 204

Tool #81 Profitability Tools.................................... 207
    Earnings Surprises 208
        Figure 77 Earnings Estimates 208
    Horizontal (Trend) Analysis 208
        Figure 78 Profit Margin in Annual Report..................... 209
    Profitability Ratios 209
        Figure 79 Return on Equity for Stocks 211
    Quality of Earnings... 212
    Vertical (Common-Size) Analysis.............................. 214

Tool #82 Real Estate Tools 216
    Capitalization Rate 216
    Home Price Statistics 216
        Figure 80 Home Sales Report 217
    Net Income Multiplier (NIM).................................. 218
    NCREIF Property Performance Averages....................... 219
        Figure 81 NCREIF's Real Estate Performance Averages 219

Tool #83 Russell 2000 Index 220
    Figure 82 Russell Indexes vs. Other Key Indexes 222

Tool #84 Safety and Timeliness Ranking 223

Tool #85 Share-Price Ratios 224
    Book Value per Share 224
    Cash per Share 225
    Earnings per Share 226
    Growth Rate 227
        Figure 83 Future Value of $1.00 Table 229
    Price-to-Book Value Ratio 230
        Figure 84 Price-to-Book Value Ratio for Stocks 231
    Price-Earnings Ratio (Multiple) 232
        Figure 85 Price-to-Earnings Ratio for Stocks 233
    Price-to-Sale Ratio (PSR) 234
        Figure 86 Price-to-Sales Ratio for Stocks 235

Tool #86 Sharpe's Risk-Adjusted Return 236

Tool #87 Short Selling 237
    Short-Interest Ratio (SIR) 237
    Short Sales Position 238
    Specialists/Public Short (SIP) Ratio 239

Tool #88 Standard & Poor's 500 Indexes 240
    Figure 87 The Standard & Poor's 500 Index 240

Tool #89 Standard & Poor's Other Stock Indexes 242
    Figure 88 Standard & Poor's Other Indexes 243

Tool #90 Stock Splits 244
    Figure 89 Stock Splits 245

Tool #91 Support and Resistance Levels 246
    Figure 90 Support and Resistance Levels Chart 247

Tool #92 Tick and Closing Tick 248

Tool #93 Total Return 249
    Figure 91 Total Return for Large Mutual Funds 251

Tool #94 Trading Volume 252

Tool #95 Trading Volume Gauges 254
Low-Price Activity Ratio 254
Net Member Buy–Sell Ratio 254
Figure 92 Trading Volume Gauges. 255
Odd-Lot Theory 256
Figure 93 Odd-Lot Trading 256
Up-to-Down Volume Ratio 257

Tool #96 Treynor's Performance Measure 258

Tool #97 Value Averaging 259

Tool #98 Value Line Averages 260
Value Line Convertible Indexes 261

Tool #99 Wilshire 5000 Equity Index 262
Figure 94 Wilshire 5000 Index 263

Tool #100 Yield Curve 264
Figure 95 U.S. Treasury Yield Curve 264

Tool #101 Yield on an Investment 266
Annual Percentage Rate (Effective Annual Yield) 266
Figure 96 Effective Interest Rates with Compounding 267
Current Yield on a Bond 267
Current Yield on a Stock 268
Figure 97 Dividend Yield for Stocks 269
Dividend Payout Ratio 270
Figure 98 Dividend Payout Ratio. 271
Tax-Equivalent Yield. 271
Figure 99 Tax Equivalent Yield 272

Online Internet Resources 273

TOOL #1

AMEX MAJOR MARKET INDEX

What Is This Tool?   The American Stock Exchange (AMEX) Major Market Index is a price weighted arithmetic average of 20 high-quality industrial New York Stock Exchange (NYSE) securities.

How Is It Computed?   The index is prepared by the AMEX but includes only stocks listed on the NYSE, 19 of which are included in the Dow Jones Industrial Average: American Express, AT&T, Chevron, Coca-Cola, Walt Disney, Dow Chemical, Du Pont, Eastman Kodak, Exxon, General Electric, General Motors, IBM, International Paper, Johnson & Johnson, McDonald's, Merck, 3M, Philip Morris, Procter & Gamble, and Sears. High priced issues have a greater impact on the index than low-priced issues. The base level of 200 was established in 1983.

FIGURE 1   Here's a look at how a web site covers the AMEX Major Market Index. (Reproduced with permission of Yahoo! Inc.© 2000 by Yahoo! Inc. YAHOO! and the YAHOO! Logo are trademarks of Yahoo! Inc.)

Where Is It Found? The index appears in online databases like Yahoo! and in the financial pages of newspapers such as *Barron's* and *The Wall Street Journal,* and investment magazines. Also, look at the AMEX Web page on the Internet at www.amex.com. (See Figure 1).

How Is It Used for Investment Decisions? This index, and trading in its related options, may be used to confirm market moves in other indexes containing blue chip stocks such as the Dow Jones Industrial Average.

A Word of Caution: This index was created in an era when the owners of the Dow Jones average of 30 industrial stocks did not let that benchmark be used for index options trading. Now that Dow Jones allows trading on the Dow 30, this index has fallen in popularity.

Also See: Dow Jones Industrial Average.

Tool #2

Amex Composite Index

What Is This Tool? The AMEX Composite is a market value-weighted index reflecting the performance of the companies whose shares trade on the American Stock Exchange.

How Is It Computed? The index reflects the relative value of all the shares of all the stocks, real estate trusts, limited partnerships, American Depository Receipts (ADRs), and closed-end funds trading on the AMEX. Five major industry-based subindexes (information technology, financial, healthcare, natural resources, and industrial) are also derived from the index calculation. The AMEX Composite had a base level of 550 as of December 29, 1995.

Where Is It Found? The index may be found in the financial section of major newspapers such as *Barron's, Investor's Business Daily,* and *The Wall Street Journal.* It can also be found in the electronic online service, America Online, or on the Internet at www.amex.com.

How Is It Used for Investment Decisions? The index is used to gauge the performance and trading activity of the stocks on the AMEX, typically seen as a home to second-tier companies. Therefore, an investor owning securities from smaller or young companies may find the index helpful in creating a benchmark for that kind of investments.

Also See: Russell 2000, Standard & Poor's Other Indexes.

Tool #3

Amex: Other Indexes

INSTITUTIONAL INDEX

What Is This Tool? This index tracks the performance of 75 companies that have significant positions held by institutions.

How Is It Computed? This is a broad capitalized market value-weighted index of the shares of companies widely held by large institutional investors. To be eligible to be listed, a minimum of 7 million shares must be traded and at least 200 institutional investors must hold the stock. As of November 1998, companies included in the index are Abbott Labs, Airtouch, Allied Signal, Allstate, American Express, American Home Products, American International Group, Ameritech, Amoco, Associated First Capital, AT&T, BankAmerica, Banc One, Bell Atlantic, BellSouth, Boeing, Bristol-Myers, Chase Manhattan, Chevron, Chrysler, Cisco, Citigroup, Coca-Cola, Compaq, Dayton Hudson, Dell, Disney, du Pont, Emerson Electric, Exxon, Fannie Mae, First Union, Ford, Freddie Mac, GE, General Motors, Gillette, GTE, Hewlett-Packard, Home Depot, Intel, IBM, Johnson & Johnson, Kellogg, Kimberly-Clark, Lilly, Lockheed Martin, Lucent, McDonald's, MCI WorldCom, Medtronic, Merck, Merrill Lynch, Microsoft, 3M, Mobil, Monsanto, Morgan Stanley Dean Witter, PepsiCo, Pfizer, Philip Morris, Procter & Gamble, Royal Dutch, SBC, Schering-Plough, Schlumberger, Sears, Texaco, Time Warner, Tyco, Wal-Mart, Warner-Lambert, Wells Fargo, and Xerox.

A base level of 25 was assigned in 1986. The list of stocks is updated on a quarterly basis when appropriate.

Where Is It Found? Information on the index is in *Barron's* on the AMEX Web page on the Internet at www.amex.com.

How Is It Used for Investment Decisions? An investor may judge the performance of actively traded securities held in large measure by institutional investors. Institutions are considered "smart money" and a trend in the index may indicate changes in institutional perceptions of the overall stock market. The individual investor may want to follow institutional leads when they appear sound.

The index may be a proxy for an investor's "core" stock holdings. Stocks with large institutional holders tend to be conservative.

Also See: Standard & Poor's 500 Index.

EUROTOP 100 INDEX

What Is This Tool?   The index measures the collective performance of the most actively traded stocks on Europe's major stock exchanges and is designed to be representative of the European stock market as a whole.

How Is It Computed?   The index comprises the 100 most actively traded stocks in 9 European countries: Belgium, France, Germany, Italy, The Netherlands, Spain, Sweden, Switzerland, and the United Kingdom. The index is weighted based on each country's total stock market capitalization and gross domestic product, a weighting that determines a country's total number of stocks as well as the number of shares of the stocks represented in the index.

The index's aggregate price is in European Currency Units (ECUs). Country weightings are based on each nation's exchange capitalization at the end of the previous calendar year.

As of November 1998, the index comprised the following stocks, categorized by their home country:

Belgium   Electrabel SA, Generale de Banque SA, Petro-Fina SA.

France   Alcatel Alsthom CGE, Axa-UAP, Carrefour Supermarche SA, Compagnie Generale des Eaux, Compagnie de Saint-Gobain, Compagnie de Suez, Elf Aquitaine SA, Group Danone, LVMH, L'Oreal, PSA Peugeot Citroen, Societe Generale, Total SA-B.

Germany   Allianz, BASF, Bayer, BMW, Commerzbank, Daimler-Benz, Deutsche Bank, Hoechst, Mannesmann, RWE, Siemens, Thyssen, VEBA, Volkswagen.

Italy   Assicurazione Generali, ENI, FIAT, Montedison, Stet Societa' Finanz. Tel., Telecom Italia.

The Netherlands   ABN Amro, Akzo Nobel, Elsevier, ING, Koninklijke PTT Nederland, Philips Electronics, Royal Dutch Petroleum, Unilever.

Spain   Banco Santander, Endesa, Iberdrola, Repsol, Telefonica de Espana.

Sweden   Astra, LM Ericsson, Investor, S.E. Banken, Telefonakicbolaget, Volvo.

Switzerland   Asea Brown Boveri Ltd., Credit Suisse Group, Nestle SA, Novartis AG, Roche Holding, Schweizerischer Bankverein, Swiss Reinsurance, Zuerich Versicherungsgesellschaft.

United Kingdom   Abbey National, Allied Domecq, B.A.T Industries, BG plc, BTR plc, Barclays, Bass, Boots, British Aerospace, British Airways, British Petroleum, British Steel, British Telecommunications, Cable & Wireless, General Electric Co. plc, Glaxo Wellcome, Granada, Grand Metropolitan, Guiness, HSBC, Imperial Chemical, Lloyds TSB, Marks & Spencer, National Westminster Bank, National Power, Prudential Corporation, Reed International, Reuters, Rio Tinto, Sainsbury, Shell Transport & Trading, SmithKline Beecham, Standard Chartered, Tesco, Unilever, Vodafone Group, Zeneca Group plc.

Where Is It Found? Information on this index can be obtained at www.amex.com and information on trading in options tied to this index runs in *The Wall Street Journal.*

How Is It Used for Investment Decisions? The index indicates how foreign company stocks are performing. An investor wishing to diversify his portfolio internationally may use this index as a way to gauge the performance of overseas company stocks that trade in U.S. markets.

Also See: Morgan Stanley EAFE Index.

INTER@ACTIVE WEEK INTERNET INDEX

What Is This Tool? The index is designed to track the stock market performance of the hottest idea in commerce in recent decades, the Internet. This quirky network of computers is revolutionizing how people communicate, shop, and even invest.

How Is It Computed? The index, developed by the AMEX and the Inter@active Week computer-industry magazine, tracks a cross section of 50 leading companies involved in providing computerized communications services, developing and market-

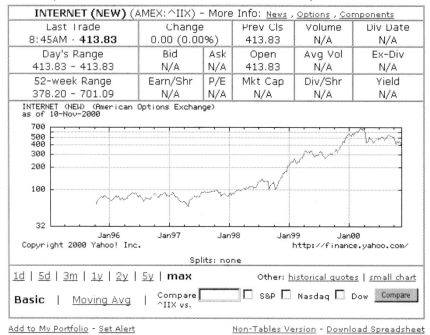

FIGURE 2 How this Internet-stock benchmark is tracked by an investment web site. (Reproduced with permission of Yahoo! Inc. © 2000 by Yahoo! Inc. YAHOO! and the YAHOO! Logo are trademarks of Yahoo! Inc.)

ing software, and making hardware for this burgeoning field. The index is market-value weighted, meaning its value is based on the sum of the share price times number of shares outstanding of each of the component stocks. (See Figure 2.)

As of December 1998, members of the index were 3Com, Adobe Systems, Amazon.com, America Online, Ascend Communications, At Home, Avid Technology, BroadBand Technologies, CINET, Cabletron Systems, C-Cube Microsystems, Check Point, CheckFree Holdings, Cisco Systems, CMG Information Services, Cybercash, Cylink, DoubleClick, E*Trade Group, EarthLink Network, Excite, Harbinger, Infoseek, Intuit, Macromedia, Mindspring Enterprises, Netscape Communications, Network Associates, Newbridge Networks, Novell, Onsale, Open Market, Pairgain Technologies, PictureTel, PSINet, Qualcomm, RealNetworks, Secure Computing, Security Dynamics Technologies, Silicon Graphics, SportsLine USA, Spyglass, Sterling Commerce, Sun Microsystems, USWeb, VeriSign, Verity, VocalTec, and Yahoo!.

Where Is It Found? While not widely quoted, this index can be found on some Internet online business news pages such as the one on Inter@active Week's Web page at www.nordby.com/clients/zd/net.asp, or at www.amex.com, and information on trading in options tied to this index runs in *The Wall Street Journal.*

How Can It Be Used for Investment Decisions? The index can tell investors which way Internet shares are headed. This is a group that tends to move in waves, so a sharp movement in the index will likely mean that Internet-related stocks will be impacted by that momentum.

A Word of Caution: Internet stocks which are promising assets are certainly not for the skittish. Some of these stocks have produced phenomenal one-day gains, only to give them back as quickly. All of that happening, at times, without a sliver of corporate news to account for the volatility.

Also See: NASDAQ Indexes.

OIL INDEX

What Is This Tool? This is an index of the stock prices of top oil companies.

How Is It Computed? The index is a capitalized-weighted market value based on the market prices per share of major companies involved in exploring, manufacturing, or developing oil products. As of December 1998, index members were Amerada Hess, Amoco, Atlantic Richfield, British Petroleum, Chevron, du Pont, Exxon, Kerr-McGee, Mobil, Occidental Petroleum, Oryx Energy, Phillips Petroleum, Royal Dutch Petroleum, Sun Company, Texaco and Unocal.

Where Is It Found? Information on this index can be obtained at www.amex.com and information on trading in options tied to this index runs in *The Wall Street Journal.*

How Is It Used for Investment Decisions? The investor may examine how high or low the index is as a basis for determining the purchase or sale of oil stocks or whether oil prices are overvalued or undervalued. The index is used as a price basis for settlement of petroleum option contracts.

Also See: Crude Oil Spot Price.

Tool #4

ARBITRAGE

What Is This Tool?   Arbitrage is the process of simultaneously buying and selling the same securities in different markets. The investor takes advantage of the price differential between the same (or comparable securities) simultaneously trading on two different exchanges. The security is bought from the exchange having the higher-priced security while, at the same time, the investor sells the security on the lower-priced exchange. Brokerage commissions have to be paid on both the "buy" and the "sell." Arbitrage takes advantage of market inefficiencies while eliminating them in the process.

How Is It Computed?   Arbitrage profit $= (Y_b - X_a) \times Q$

where

> Y_b = Price of higher-priced comparable security on Exchange B
> X_a = Price of lower-priced comparable security on Exchange A
> Q = Quantity

Example: Stock in RDL & Co. is trading on the NYSE for $4 per share and simultaneously trading on the London Exchange for $4.30 per share. An investor buys 10,000 shares of RDL stock on the NYSE and at the same time sells 10,000 shares on the London Exchange. The profit is

$$\text{Arbitrage profit} = (\$4.30 - \$4) \times 10,000 = \$3000$$

The arbitrage profit is $3000 in this transaction. This transaction and similar transactions would increase the demand and therefore the price of the NYSE security while simultaneously lowering the price of the NYSE security traded on the London Exchange. This would continue until the prices of the two securities were in parity.

Where Is It Found?   An investor can find the prices of a security on different exchanges by using a personal computer and telecommunications software to access an online database of price quotations. When price differences are noted, they must be taken advantage of instantly.

How Is It Used for Investment Decisions?   The investor uses arbitrage when seeking to exploit the price variation between the same or comparable securities or commodities on two different exchanges. These opportunities exist for fleeting periods of time, and the investor must act quickly in order to take advantage of these differences. Another arbitrage strategy is to purchase stock in a company soon to be taken over (in a deal paid in shares of the buyer) and sell the stock of the acquiring company.

A Word of Caution: Arbitrage can result in losses if a sudden, unexpected adverse price occurs in the stock of companies involved in this tactic.

Also See: Index Arbitrage.

Tool #5

Arms Index (TRIN)

What Is This Tool?   The Arms Index (TRIN for short), developed by Richard W. Arms, Jr., is a short-term trading index that offers the day trader as well as the long-term investor a look at how volume — not time — governs stock price changes. It is also commonly referred to by its quote machine ticker symbols, TRIN and MKDS.

The Arms Index is designed to measure the relative strength of the volume associated with advancing stocks vs. that of declining stocks. If more volume goes into advancing stocks than declining stocks, the Arms' Index will fall to a low level under 1.00. Alternatively, if more volume flows into declining stocks than advancing stocks, the Arms Index will rise to a high level over 1.00.

It helps to forecast the price changes of market indexes as well as of individual issues. You will find Arms indices for the NYSE, the NASDAQ market, the AMEX, and Giant Arms (a combined index for NASDAQ and AMEX). There is also the Bond Arms Index which helps forecast interest rates.

How Is It Computed?   The Arms Index is calculated by dividing the ratio of the number of advancing issues to the number of declining issues by the ratio of the volume of advancing issues to the volume of declining issues. It is computed separately for the NYSE, the American Stock Exchange, and NASDAQ. (See Figure 3.)

Using the data in the figures given in Diaries of *The Wall Street Journal:*

$$\frac{\dfrac{\text{Advances}}{\text{Declines}}}{\dfrac{\text{Advance Volume}}{\text{Decline Volume}}} = \frac{1{,}130/775}{166{,}587/78{,}016} = 0.70$$

FIGURE 3   Computing an Arms Index

Where Is It Found?   It is found in *Barron's* and *The Wall Street Journal* and reported daily on TV networks such as *Nightly Business Report, CNNfn,* and *CNBC.* (See Figure 4).

How Is It Used for Investment Decisions?   A figure of less than 1.0 indicates money flowing into stocks (bullish sign), while a ratio of greater than 1.0 shows money flowing out of stocks (bearish sign).

ARMS INDEX

The Arms index, also known as the short term
trading index, is the average volume of declining
issues divided by the average volume of advancing
issues. It is computed separately for the NYSE, the
American Stock Exchange and Nasdaq. A figure of
less than 1.0 indicates more action in rising stocks.

Daily	Aug. 21	22	23	24	25
NYSE	.82	.86	1.06	1.03	.90
AMEX	.25	1.84	.24	.22	2.53
NASDAQ	.66	.67	.53	.65	1.23

FIGURE 4   The Arms Index. How a newspaper covers the Arms Index. (From *Barron's Market Week,* August 28, 2000, p. MW72. With permission.)

One variation of the Arms Index that many technicians monitor is the Open 10 TRIN (also known as the Open 10 Trading Index). It is calculated by

Taking a ratio of a 10-day total for the number of advancing issues to a 10-day total of the number of declining issues and dividing it by a ratio of 10-day total of the volume of advancing issues to a 10-day total of the volume of declining issues.

In addition, a 30-day version of the Open 10 TRIN is frequently used.

High readings reflect an oversold condition and are generally considered bullish. Low readings reflect an overbought condition and are generally deemed bearish.

A Word of Caution: Many studies have been performed on the Arms Index with various conclusions. Many indicate that the Arms Index has relatively limited forecasting value for stock prices.

Also See: Tick and TRIN.

Tool #6

Asset Allocation

What Is This Tool?   Asset allocation measures the weighing of various types of investments in a portfolio. The changing of asset allocations is an attempt to maximize return while minimizing risks. The calculation can be applied to professionally managed portfolios as well as to an individual's holdings. The most widely discussed asset allocation is some combination of stocks, bonds, and cash although other asset types can be used.

How Is It Computed?   The mathematics can be simple. To determine his or her asset allocation mix, an investor can add his or her holdings of stocks, bonds, and cash, and divide each sum by the total value of the portfolio.

However, in today's complex investment world, determining what asset class certain investments belong to can be confusing. An investor will have to give some thought to how he or she allocates mixed investments such as balanced mutual funds that own both stocks and bonds or how her or she treats convertible securities which are half-bond and half-stock.

The calculation can be done by hand or by using a spreadsheet software program like *Microsoft Excel* with the help of personal finance software such as *Quicken* or *Microsoft Money*. Mutual fund owners who have access to the Internet may want to use *Morningstar Inc.'s* www.morningstar.com Web site to calculate the asset allocation from their funds. (See Figures 5 and 6.)

Where Is It Found?   Most major brokerages maintain a recommended asset allocation that is updated to keep up with the investment climate. Each quarter *The Wall Street Journal* tracks what Wall Street firms are suggesting and how the various brokerages' allocation recommendations have performed previously. In addition, many market newsletters and money management firms also tell investors what they believe are good allocations for the times. Asset allocation mutual funds own a mix of stocks, bonds, and cash, and allow investors to have a professional manager adjust the ratios of those holdings as market conditions warrant.

How Is It Used for Investment Decisions?   Many professionals believe that asset allocation is among the most important decisions for investors, and many theories are applied to asset allocation selection.

One popular theory is pegged to an investor's time horizon. Longer-term portfolios can allow an investor to take on, with some comfort, more risk. That means he or she can increase the share allocated to stocks or real estate or other volatile investments. Changes in allocation to lower-risk assets would be likely to be made as the individual's time horizon narrows or as profits or losses change the portfolio's composition.

Asset	Dollar Value of Asset (A)	Percent of Asset in Stocks (B)	Percent of Asset in Bonds [C]	Percent of Asset in Cash (D)	Dollars in Stocks (A × B)	Dollars in Bonds (A × C)	Dollars in Cash (A × D)
	$	%	%	%	$	$	$
	$	%	%	%	$	$	$
	$	%	%	%	$	$	$
	$	%	%	%	$	$	$
	$	%	%	%	$	$	$
	$	%	%	%	$	$	$
	$	%	%	%	$	$	$
	$	%	%	%	$	$	$
	$	%	%	%	$	$	$
	$	%	%	%	$	$	$
	$	%	%	%	$	$	$
	$	%	%	%	$	$	$
	$	%	%	%	$	$	$
	Total portfolio value (G)				Total stock value (H)	Total bond value (I)	Total cash value (J)
Sum values in columns	$				$	$	$
					Share in stocks (H ÷ G)	Share in bonds (I ÷ by G)	Share in cash (J ÷ by G)
Divide as directed to get your asset allocation					%	%	%

FIGURE 5   Asset Allocation Grids. Here is a handy guide to calculating your own asset allocation. You can do this by hand, or recreate this layout on a computer spreadsheet program.

A Word of Caution: Another tactic is to try to time changes in asset allocation to investment market changes, hopefully, in step with market cycles. In bull markets, investors would want to be heavily into stocks, for example. When stocks are out of favor, cash, bond, or precious metal allocations would be high. There is great debate whether individuals or professionals can profitably time market swings.

Stocks (40%)
  Equity Growth Fund (20%)
  Value Fund (20%)

Bonds (45%)
  Bond Fund (40%)
  International Bond Fund (5%)

Money Market (15%)
  Capital Preservation Fund

Moderate Portfolio

Stocks (60%)
  Equity Growth Fund (20%)
  Heritage Fund (20%)
  Ultra Fund (20%)

Bonds (30%)
  GNMA Fund (15%)
  Intermediate-Term Bond Fund (15%)

Money Market (10%)
  Capital Preservation Fund

Aggressive Portfolio

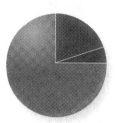

Stocks (75%)
  International Growth Fund (10%)
  Growth Fund (10%)
  Small Cap Value Fund (15%)
  Ultra Fund (25%)
  Vista Fund (15%)

Bonds (20%)
  Intermediate-Term Bond Fund (15%)
  International Bond Fund (5%)

Money Market (5%)
  Capital Preservation Fund

FIGURE 6A   Here's how some firms suggest investors divide up their savings by asset category. (Courtesy of American Century Investments, "Investing with a Purpose," 2000.)

Choosing an asset allocation strategy according to your score*

Your Score	Suggested Mix of Asset Classes**	Average Annual Return (1926–1999)	Worst Annual Loss (1926–1999)	Number of Years With a Loss (1926–1999)	Types of Funds for Your Portfolio
69–75	100%	11.3%	–43.1%	20 of 74	**Stock funds** Actively managed growth or value funds, or index funds that track the total stock market or just a segment of the market, such as the S&P 500 Index.
59–68	20% / 80%	10.5%	–34.9%	19 of 74	**Balanced funds** Actively managed or index funds that hold a mix of stocks, bonds, and (sometimes) cash investments. This type of "all-in-one" fund can automatically maintain your target asset allocation through a single investment. A balanced fund can also be designed as a "fund of funds," meaning it invests in other mutual funds rather than buying individual stocks and bonds directly.
51–58	40% / 60%	9.5%	–26.6%	17 of 74	**Bond funds** Actively managed short-, intermediate-, or long-term corporate, government, or tax-exempt funds, or comparable bond index funds.
40–50	50% / 50%	8.9%	–22.5%	16 of 74	**Money market funds** Actively managed taxable or tax-exempt funds that invest in "money market instruments" (also known as cash investments) that are issued by governments, corporations, banks, or other financial institutions.
32–39	40% / 60%	8.3%	–18.4%	16 of 74	*These are sample allocations only. Depending on your tolerance for risk or your individual circumstances, you may wish to choose an allocation that is more conservative or aggressive than the model suggested by your score. Keep in mind that these portfolios are for specific, longer-term financial goals. You may very well hold cash investments for shorter-term goals and emergencies.
23–31	20% / 80%	7.0%	–10.1%	13 of 74	**Cash investments are represented by U.S. Treasury bills, bonds by long-term U.S. corporate bonds, and stocks by the Standard & Poor's 500 Composite Stock Price Index (S&P 500 Index). Returns include the reinvestment of all income, dividend, and capital gains distributions.
11–22	10% / 10% / 80%	6.1%	–6.7%	10 of 74	Source: The Vanguard Group.

░ **Stocks**   ▓ **Bonds**   ▓ **Cash Investments**

FIGURE 6B   ©1999 The Vanguard Group, Inc. Reprinted with permission from the "Plain Talk" brochure, *The Vanguard Investment Planner*. All rights reserved. Vanguard Marketing Corporation, Distributor.

How a Major Wall Street Firm Advises Clients on Asset Allocation as of October, 2000

Objective	Allocation		
Capital Preservation	10% stocks	55% bonds	35% cash
Income	30% stocks	60% bonds	10% cash
Income & Growth	40% stocks	50% bonds	10% cash
Growth	70% stocks	25% bonds	5% cash
Aggressive Growth	80% stocks	10% bonds	10% cash

FIGURE 6C   (From Merrill Lynch & Co. With permission.)

Tool #7

BARRON'S INDEXES

CONFIDENCE INDEX

What Is This Tool? The index reflects the trading pattern of bond investors to determine the timing of buying or selling stock. It is generally assumed that bond traders are more knowledgeable than are stock traders. Consequently, bond traders identify stock market trends sooner. The index is used in technical investment analysis.

How Is It Computed? *Barron's* Confidence Index equals

$$\frac{\text{Yield on Barron's 10 top-grade corporate bonds}}{\text{Yield on Dow Jones 40 bond average}}$$

The numerator will have a lower yield than the denominator because it consists of higher-quality bonds (rated AAA or AA). Some high-quality bonds are those of AT&T, General Electric, and Procter & Gamble. The lower the risk of default, the lower the return rate is.

Assume the Dow Jones yield is 8 percent and the *Barron's* yield is 7 percent. The Confidence Index is

$$\frac{7\%}{8\%} = .875 = 87.5\%$$

Where Is It Found? The index is published weekly in *Barron's*.

How Is It Used and Applied? Because top-quality bonds have lower yields than lower-grade bonds, the index will be below 100 percent. Typically, the trading range is between 80 and 95 percent. When bond investors are bullish, yield differences between the high-grade bonds and low-grade bonds will be small. In such cases, the index may be close to 95 percent.

If the feeling is bearish, bond market investors will want to hold top-quality issues. Some investors who continue to put their money in average-quality bonds or lower-quality bonds will want a high yield for the increased risk. The Confidence Index will then decline because the denominator will be getting larger. If confidence is high, investors are apt to purchase lower-grade bonds. As a result, the yield on high-grade bonds will decrease while the yield on low-grade bonds will increase.

How Is It Used for Investment Decisions? When bond traders are ignoring lower-quality bonds, it can be either a signal that the economy is in trouble or that a buying opportunity for those lower-quality bonds exists.

If an investor knows what bond traders are doing now, he or she also may be able to predict what stock traders will be doing in the future. The lead time between the Confidence Index and economic conditions and stock market performance is considered to be several months. This leaves ample time for an investment decision.

If bond traders are bullish, an investor may invest in stocks now before stock prices rise. On the other hand, if bond traders are bearish, an investor would not buy stocks or consider selling current holdings on the expectation that stock prices will fall.

A Word of Caution: Bond traders may be making the wrong investment decision which could result in misleading inferences about stock prices. The Confidence Index has a mixed track record in predicting the future. The index is deficient in that it considers investors' attitudes only on yields (simply, a look at the demand function). It ignores the supply of new bond issues (supply pattern) as they affect yields. A large bond issue by a major corporation, for example, may result in increased high-grade bond yields. Such movement would occur despite heavy demand for single issues when prevailing investor attitudes believe that yields should be dropping.

50-STOCK AVERAGE

What Is This Tool?   *Barron's* 50-Stock Average is an unweighted average of 50 leading NYSE listed issues.

How Is It Computed?   Each stock is given equal weight in determining the average. It is an unweighted price average—calculated weekly—that includes both earnings and dividends of the 50 issues, and assumes that the same dollar investment is made in each component company instead of one share of each company.

Where Is It Found?   It is found in *Barron's*. (See Figure 7.)

How Is It Used for Investment Decisions?   The main value of this index is the analysis that *Barron's* provides. Thus, each week, an investor can see comparisons to projected quarterly and annual earnings. This average is intended to be used as a yardstick in deciding whether the market, at any given time, is reasonably priced, overvalued, or undervalued in view of what the investor believes to be the most probable course of the business cycle.

A Word of Caution: Like the Dow Jones Industrial Average, it is not a broad market index. It should not be interpreted as such.

Also See: AMEX Major Market Index, Dow Jones Industrial Average, Standard & Poor's 500 Index.

BARRON'S 50-STOCK AVERAGE

This 50-stock index is an unweighted average of 50 leading issues with each stock given equal weight in determining the average. It offers comparisons to the projected quarterly and annual earnings which appear in the table. The earnings yield, which is the reciprocal of the Price/Earnings Ratio (1 divided by the P/E), can be compared to bond yields in the table. The dividend yield equals the dividend divided by the price of the average.

	Aug. 24 2000	Aug. 17 2000	Aug. 1999
Average price index	4141	4132	4341
Projected quarterly earn	54.86	54.86	56.20
Annualized projected earn	219.44	219.44	224.80
Annualized projected P/E	18.9	18.8	19.4
Five-year average earn	181.96	181.96	173.38
Five-year average P/E	22.8	22.7	25.0
Year-end earn	204.27	204.27	177.32
Year-end P/E	20.3	20.2	24.5
Year-end earns yield, %	4.9	4.9	4.1
Best grade bond yields, %	7.31	7.38	7.29
Bond yields/stock ylds, %	1.49	1.51	1.79
Actual year-end divs	77.76	77.76	75.44
Actual yr-end divs yld, %	1.88	1.88	1.74

FIGURE 7   *Barron's* 50-Stock Index. How *Barron's* covers its 50-stock index. (From *Barron's Market Week*, August 28, 2000, p. MW 76. With permission.)

TOOL #8

BETA FOR A MUTUAL FUND

What Is This Tool?   Beta is a measure of uncontrollable risk that results from forces outside of the mutual fund's control. Purchasing power, interest rate, and market risks fall into this category. This type of risk is measured by *beta*.

How Is It Computed?   In measuring a fund's beta, an indication is needed of the relationship between the fund's return and the market return (such as the return on the Standard & Poor's 500 Stock Composite Index). This relationship is statistically computed, which is not covered here.

Where Is It Found?   Betas for mutual (stock) funds are widely available in many investment newsletters and directories. Examples are *S&P's Stock Guide, Money Magazine, Business Week, Forbes, Fortune, The Individual Investor's Guide to No-load Mutual Funds,* and various reports from Morningstar Inc. and Lipper Analytical Service. Financial advisory services such as *Morningstar* (Internet users can check out www.morningstar.net), *Micropal,* and *Lipper Analytical Service* also provide their own risk-adjusted rating systems developed, based on beta and return.

How Is It Used and Applied?   Beta measures a stock fund's volatility relative to an average fund. In assessing the risk or instability of a mutual fund, beta is widely used. Beta shows how volatile a mutual fund is compared with the overall market, typically measured by the Standard & Poor's 500 index of the most widely held stocks. For example, if the S&P goes up 10 percent and your fund goes up 10 percent in the same period, the fund has a beta of 1. However, if the fund goes up 20 percent, it has a beta of 2, meaning it is twice as volatile as the market. The higher the beta, the greater the risk.

Beta	What It Means
1.0	A fund moves up and down just as much as the market.
>1.0	The fund tends to climb higher in bull markets and dip lower in bear markets than the S&P index.
<1.0	The fund is less volatile (risky) than the market.

How Is It Used for Investment Decisions?   Beta of a particular fund is useful in predicting how much the fund will go up or down, provided that investors know the direction of the market. Beta helps to figure out risk and expected (required) return. (See Figure 8.)

Expected (required) return = risk-free rate + beta × (market return − risk-free rate)

The higher the beta for a fund, the greater the return expected (or demanded) by the investor.

Fund Name	Beta	5-Year Average Total Return
Fidelity Magellan	0.97	14.7%
Vanguard Index 500	1.00	19.8%
Washington Mutual Investors	0.83	18.9%
Investment Comp of America	0.86	17.0%
Fidelity Growth & Income	0.91	18.6%
Fidelity Contrafund	0.90	16.6%
Vanguard/Windsor II	0.91	18.0%
Vanguard/Wellington	0.62	15.0%
Fidelity Puritan	0.65	13.2%
American Century (20th) Ultra	1.26	14.6%
Income Fund of America	0.49	12.7%
Fidelity Advisors Growth Opportunity	0.85	17.7%
EuroPacific Growth	0.72	9.9%
Fidelity Equity-Income	0.92	15.8%
Vanguard/Windsor	1.02	12.4%
New Perspective	0.79	14.2%
Vanguard Institutional Index	1.00	19.9%
Putnam Growth & Income A	0.89	15.8%
Janus Fund	0.97	15.7%
Pimco Total Return	1.19	8.0%
Fidelity Equity-Income II	0.91	15.6%
MSDW Dividend Growth B	0.82	15.7%
Putnam Growth & Income B	0.89	14.9%
Fidelity Blue Chip Growth	1.00	18.1%
Franklin California Tax-Free Income	0.64	6.3%
Growth Fund of America	1.00	14.9%
Putnam Voyager A	1.13	15.5%
Templeton Foreign I	0.69	6.0%
Janus Worldwide	0.74	17.8%
Templeton Growth I	0.77	10.5%
Fundamental Investors	0.85	16.7%
Fidelity Spartan U.S. Index	1.00	19.6%
AIM Constellation A	1.17	12.0%
T. Rowe Price Equity-Income	0.68	17.0%
Fidelity Asset Manager	0.63	10.7%
Vanguard U.S. Growth	0.98	21.5%
IDS New Dimensions A	1.02	16.4%
Vanguard/Primecap	1.00	19.8%
American Mutual	0.65	14.6%
Vanguard GNMA	0.75	7.4%
Putnam New Opportunities A	1.25	15.7%
Fidelity Growth Company	1.10	15.2%
Templeton World I	0.79	12.4%
Bond Fund of America	0.61	6.4%
T. Rowe Price Int'l Stock	0.67	7.7%
Janus Twenty	1.11	22.3%
Franklin U.S. Government Securities I	0.66	6.8%
Capital Income Builder	0.51	13.6%
Smallcap World	0.86	8.1%
Fidelity Low-Priced Stock	0.73	15.1%

FIGURE 8   Betas for Large Mutual Funds. Here is a look at betas and 5-year average total return for the 50 largest mutual funds. All data is as of Sept. 30, 1998. (From Morningstar Inc. With permission.)

Example: Gee Whiz Fund actually returned 25 percent. Assume that the risk-free rate (for example, return on a T-bill) equals a 6 percent market return (for example, return on the S&P 500) equals 12 percent, and Gee Whiz Fund's beta equals 2. Then, the return on the Gee Whiz Fund required by investors would be

$$\text{Expected (required) return} = 6\% + 2\,(12\% - 6\%)$$
$$= 6\% + 12\%$$
$$= 18\%$$

Because the actual return (25 percent) is significantly above the required return (18 percent), you would be willing to buy the shares of the fund.

A Word of Caution: First, you should cover at least three years of beta to get the idea about the risk and instability of the fund and, second, you should consider beta along with other selection criteria such as management fees, alpha, R^2, and standard deviation.

Also See: Alpha for a mutual fund, Lipper Mutual Fund Rankings, net asset value for a mutual fund, risk-adjusted return, R^2 for a mutual fund, standard deviation for a mutual fund.

TOOL #9

BETA FOR A SECURITY

What Is This Tool?   Many investors hold more than one financial asset. The portion of a security's risk (called unsystematic risk) can be controlled through diversification. This type of risk is unique to a given security. Nondiversifiable risk, more commonly referred to as systematic risk, results from forces outside of the firm's control and are, therefore, not unique to the given security. Purchasing power, interest rate, and market risks fall into this category. This type of risk is measured by *beta.*

How Is It Computed?   In measuring an asset's beta, an indication is needed of the relationship between the asset's return and the market return (such as the return on the Standard & Poor's 500 Stock Composite Index). This relationship is statistically computed, which is beyond the scope of this book.

Where Is It Found?   Betas for stocks are widely available in many investment newsletters and directories. Examples are *Value Line Investment Survey* and *S&P's Stock Guide.* Figure 9 shows a list of some selected betas.

Stocks	Betas
Philip Morris	1.20
Charles Schwab	2.20
Dow Chemicals	1.00
Exxon	0.65
Pfizer	1.10
General Motors	1.15
Coca-Cola	1.10

FIGURE 9   Selected Betas for Stocks. Data taken from *Value Line Investment Survey,* April–May 1997, published by Value Line, Inc., 711 3rd Avenue, New York, NY 10017.

How Is It Used for Investment Decisions?

1. Beta measures a security's volatility relative to an average security. Put it another way, it is a measure of a security's return over time to that of the overall market.
2. For example, if a mythical stock, XYZ, has a beta of 2.0, it means that if the stock market goes up 10 percent, XYZ's common stock goes up 20 percent; if the market goes down 10 percent, XYZ goes down 20 percent.
3. Here is a guide for how to read betas.

Beta	What It Means
0	The security's return is independent of the market. An example is a risk-free security, such as a T-bill.
0.5	The security is only half as responsive as the market.
1.0	The security has the same response or risk as the market (i.e., average risk). This is the beta value of the market portfolio, such as Standard & Poor's 500.
2.0	The security is twice as responsive, or risky, as the market.

The beta of a particular stock is useful in predicting how much the security will go up or down, provided that investors know which way the market will go. Beta helps to figure out risk and expected (required) return.

Expected (required) return = risk-free rate + beta × (market return − risk-free rate)

The higher the beta for a security, the greater the return expected (or demanded) by the investor.

Example: Johnstown Metals Co. stock actually returned 8 percent. Assume that the risk-free rate (for example, return on a T-bill) was 5.5 percent and the market return (for example, return on the S&P 500) was equal to 9 percent, and Johnstown Metals Co.'s beta was equal to 1.2. Then, the return on Johnstown Metals Co. stock required by investors would be

$$\text{Expected (required) return} = 5.5\% + 1.2\,(9\% - 5.5\%)$$
$$= 5.5\% + 4.2\%$$
$$= 9.7\%$$

Because the actual return (8 percent) is less than the required return (9.7 percent), you would not be willing to buy the Johnstown Metals Co. stock.

A Word of Caution: An investor should look at beta for several years to determine the stability of the security. Beta should be considered along with other selection criteria such as dividend yield, earnings growth, and price-earnings (P/E) ratio.

Also See: Beta for a Mutual Fund, Dividend Yield, Earnings Growth, Investor's *Business Daily's Smart Select*™ Tables, Price–Earnings (P/E) Ratio.

Tool #10

Bond Market Indexes

What Are These Tools? Bond market performance is tracked by a horde of indexes, none of which comes close to carrying the clout of stock market benchmarks such as the Dow Jones Industrial Average or the Standard & Poor's 500.

While many investors simply look at bonds for their yield (the income stream) because they do not intend to trade the bonds and, thus, carefully watch benchmarks like Treasury bond yields, others looked to boost their take from bonds by actively trading these securities. This is where indexes become so important as a measure of a trader's success.

The bond market is filled with large niches where investors like to specialize from government bonds to corporate issues to foreign debts. Some indexes carefully monitor those small categories while others, so-called "aggregate" indexes, take a broader view.

The Salomon Smith Barney Brothers Broad Investment-Grade Bond Index, or "BIG," is considered by many to be the key benchmark to gauge a money manager's performance in the bond market. It is a comprehensive index of bond prices for securities that have a minimum value of $25 million. BIG reflects the total return rate earned on corporate, mortgage, and Treasury securities with a maturity of one year or more. The index includes about 3750 mortgages and bonds predominantly comprised of corporate issues. The index is updated monthly for changes in its component issues and revised bond ratings. (See Figure 10).

BOND AGGREGATE INDEXES

Other bond market aggregate indexes include

> **Lehman Brothers Aggregate Bond Index** — Benchmark for investment-grade fixed-rate debt issues, including government, corporate, asset-backed, and mortgage-backed securities, with maturities of at least one year.
>
> **Lehman Brothers 1–3 Year Government/Corporate Bond Index** — Tracks government and corporate fixed-rate debt issues with maturities between one and three years.
>
> **Lehman Brothers Intermediate Government/Corporate Bond Index** — Benchmark for government and corporate fixed-rate debt issues with maturities between 1 and 10 years.

Selected Bond Market Indexes

Index	5-Year Total Return	10-Year Total Return
First Boston Convertible Bond	8.4%	11.0%
First Boston High Yield	8.5%	10.7%
JP Morgan World Government Bond	7.5%	9.2%
Lehman 1–3 Year Government Bond	5.8%	7.4%
Lehman Aggregate Bond	7.2%	9.3%
Lehman Corporate Bond	7.6%	9.9%
Lehman Government Bond	7.1%	9.3%
Lehman Intermediate-Term Treasury	6.3%	8.3%
Lehman Long-Term Treasury Bond	9.2%	11.8%
Lehman Mortgage Backed Bond	7.3%	9.1%
Lehman Municipal Bond	6.4%	8.4%
Salomon World (no U.S.) Government Bond	7.2%	9.1%
6-month CDs	5.3%	5.8%
Dow Jones Utility Index	10.1%	12.0%

FIGURE 10   Here is a look at how a selection of bond market indexes has performed on a total return bases (dividends plus price appreciation) for 5 years and 10 years ending on September 30, 1998, and how they compare to 6-month certificates of deposit and the Dow Jones Utility Index in the same periods. (From Morningstar Inc. With permission.)

GOVERNMENT BOND INDEXES

When watching U.S. government bonds, most attention is paid to trading in U.S. Treasuries. One popular index that tracks this niche is the Lehman Brothers Treasury Bond Index. This index has two components, the Intermediate Treasury Index and the Long Treasury Index. The former consists of U.S. Treasury issues with maturities of less than 10 years, while the latter is comprised of issues with maturities of 10 years or more.

Other indexes that track Treasury bonds as well as other government-backed bonds including mortgage securities are

> **Lehman Brothers Mortgage-Backed Securities Index** — Provides performance coverage of 15- and 30-year fixed-rate securities backed by mortgage pools of the Government National Mortgage Association (GNMA or "Ginnie Mae"), Fannie Mae, and Freddie Mac.
>
> **Lehman Brothers 1–5 Year U.S. government Bond Index** — Benchmark for government fixed-rate debt issues with maturities between one and five years.
>
> **Lehman Brothers Government Bond Index** — Benchmark for U.S. government and government agency securities (other than mortgage securities) with maturities of one year or more.
>
> **Ryan Treasury Index** — Tracks total return of the most recently auctioned Treasury notes and bonds with maturities of 2 to 30 years. It was developed by the

Ryan Financial Strategies Group and is made up of the current 2-, 3-, 4-, 5-, 7-, 10-, and 30-year issues that are auctioned by the Treasury on a periodic schedule. **Salomon Smith Barney 3-Month T-Bill Index** — Tracks the average of T-bill rates for each of the prior three months, adjusted to a bond equivalent basis. **Salomon Smith Barney Treasury/Agency Index** — Provides performance coverage of U.S. Treasury and U.S. government agency securities with fixed-rate coupons and weighted average lives of at least one year. **Salomon Smith Barney GNMA Index** — Provides performance coverage of 15- and 30-year fixed-rate securities backed by mortgage pools of the GNMA. **Salomon Smith Barney Mortgage Index** — Provides performance coverage of 15- and 30-year fixed-rate securities backed by mortgage pools of the GNMA, Fannie Mae, and Freddie Mac. **Standard & Poor's Government Bond Indexes and Averages** — Tracks U.S. Treasury bonds in terms of yield and price. Long-term issues for the yield average and price index are 10 and 15 years, respectively. Intermediate-term issues for the yield average and price index are 6 to 9 years and 7.5 years, respectively. Short-term issues for the yield average and price index are 2 to 4 years and 3.5 years, respectively. The yield average is the arithmetic average of four typical issues.

CORPORATE BOND INDEXES

Corporate bonds are often tracked by watching the Dow Jones Bond Average. This is an average of bond prices for major industrial companies and utilities. The bond average is based on 20 bonds and represents an equal number of industrials and utilities. There is also a separate average for each of the industrial and utility bonds (10 each). These indexes are a simple arithmetic average based on ending market prices of the bonds. Some utilities included in the average are Consolidated Edison, Philadelphia Electric, and BellSouth. Some industrials are AT&T, Bethlehem Steel, and IBM.

Other corporate bond benchmarks are

First Boston Convertible Securities Index — Provides performance coverage of over 250 convertible bonds and preferred stocks rated B-minus or above that have original par value of at least $50 million and preferred stocks must have a minimum of 500,000 shares outstanding. The index also includes U.S. dollar-denominated Eurobonds issued by U.S. domiciled companies.

Lehman Brothers Intermediate Corporate Bond Index — Provides performance coverage of investment-quality corporate bonds with maturities of 1 to 10 years.

Lehman Brothers Long-Term Corporate Bond Index — Tracks investment-grade U.S. corporate bonds with maturities greater than 10 years.

Merrill Lynch All Convertible Securities Index — Follows corporate convertible securities that must be convertible only to common stock and have a market value or original par value of at least $50 million.

Merrill Lynch High Yield Master Index — Tracks all domestic and Yankee high-yield bonds. Issues included in the index have maturities of at least one year and have a credit rating lower than BBB–Baa3 but are not in default.
Moody's Corporate Bond Index — Tracks on a total return, price-weighted basis more than 75 nonconvertible, coupon corporate bonds. These taxable bonds mature in no more than five years.
Standard & Poor's Junk Bond Indexes — Tracks high-risk, high-return corporate bonds. They are considered poor quality with substantial gain potential. There are two Standard & Poor's indexes of junk bonds—BB-rated and B-rated.

MUNICIPAL BOND INDEXES

Two widely watched municipal bond benchmarks come from *The Bond Buyer,* a daily newspaper that principally covers the municipal bond market. It publishes several benchmark averages for the municipal bond market, the two best known indexes are its "40" and "20" indexes.

The Bond Buyer 40 consists of 40 actively traded, higher-rated general obligation, municipal bonds of varying maturities. Each bond's price is converted through a complex formula to find at what price the bond would yield 8 percent. The 40 converted prices are then averaged and put through another "conversion" designed to account for changes made in the index over the years.

The Bond Buyer 20 is an index that tracks the average prices of 20 higher-rated municipal bonds that have 20-year maturities.

Other municipal indexes include

Dow Jones Municipal Bond Yield Average — Shows the weighted-average yield on tax-free municipal bonds. This is an average of the yields of low-coupon bonds in 5 states and 15 major cities.
Lehman Brothers Municipal Bond Index — Provides performance coverage of investment-grade municipal bonds with maturities of one year or more.
Lehman Brothers Insured Municipal Bond Index   Total return performance benchmark for municipal bonds that are backed by insurers with top-shelf, triple-A ratings and have maturities of at least one year.
Lehman Brothers California Municipal Bond Index — Market capitalization-weighted index of California investment-grade bonds with maturities of one year or more.
Lehman Brothers New York Insured Municipal Bond Index — Total return performance benchmark for New York investment-grade municipal bonds with maturities of at least one year.
Lehman Brothers Massachusetts Enhanced Municipal Bond Index — Index of Massachusetts investment-grade municipal bonds with maturities of one year or more.
Standard & Poor's Municipal Bond Price Index — Based on high quality (AAA to A) municipal bonds having a maturity of about 20 years. Yield to

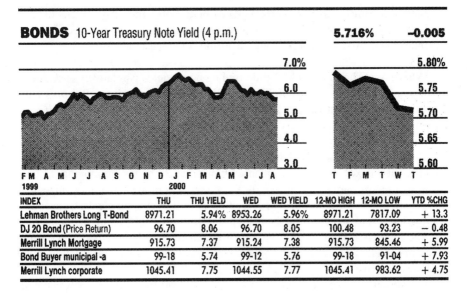

BONDS 10-Year Treasury Note Yield (4 p.m.)   **5.716%**   **–0.005**

INDEX	THU	THU YIELD	WED	WED YIELD	12-MO HIGH	12-MO LOW	YTD %CHG
Lehman Brothers Long T-Bond	8971.21	5.94%	8953.26	5.96%	8971.21	7817.09	+ 13.3
DJ 20 Bond (Price Return)	96.70	8.06	96.70	8.05	100.48	93.23	– 0.48
Merrill Lynch Mortgage	915.73	7.37	915.24	7.38	915.73	845.46	+ 5.99
Bond Buyer municipal -a	99-18	5.74	99-12	5.76	99-18	91-04	+ 7.93
Merrill Lynch corporate	1045.41	7.75	1044.55	7.77	1045.41	983.62	+ 4.75

FIGURE 11   Bond Indexes. How a newspaper covers various bond indexes. (From *The Wall Street Journal,* August 24, 2000, p. C1. With permission.)

maturity is translated to an equivalent selling price to a 20-year, 4 percent bond. An average yield is then determined for the bonds.

Standard & Poor's Municipal Bond Yield Index— Determined from 15 highly rated (AAA to A) municipal bonds. The index is an arithmetic average of the effective interest rate.

(See Figure 11).

GLOBAL BOND INDEXES

Today, investors do not have to settle just for the U.S. bond market to get fixed-income results.

The JP Morgan Government Bond Index is considered to be the most widely used benchmark for measuring performance and quantifying risk across international fixed income bond markets. The index and its underlying subindexes measure the total, principal, and interest returns in each of 13 countries and can be reported in 19 different currencies.

The index limits inclusion to markets and issues that are available to international investors to provide a more realistic measure of market performance.

Other global bond indexes include

JP Morgan Emerging Markets Bond Index Plus — Total return index of U.S. dollar and other external currency denominated Brady bonds, loans, Eurobonds, and local market debt instruments traded in emerging markets.

Salomon Smith Barney World Government Bond Index — Tracks debt issues traded in 14 world government bond markets. Issues included in the Index have fixed-rate coupons and maturities of at least one year.

Where Are They Found?   While finding bond yields is relatively easy, locating bond index results can be trickier. *The Wall Street Journal* and *Barron's* have extensive bond index coverage. Business news cable TV channels like *CNBC* and *CNNfn* also track these indexes. Internet users can check web sites such as www.bloomberg.com or www.bondsonline.com for bond index results.

How Are They Used for Investment Decisions?   Investors can look to these bond market indexes for a measurement of how prices of bonds are moving.

Rising indexes are a bullish sign for those investors motivated to maximize their total return, that is, both principal appreciation as well as coupon yields. Conversely, a falling index may indicate that the bond's tax-free status may not be worth the risk.

The municipal bond market is very fragmented with hundreds of very small issues, many with peculiar characteristics. Thus, analysts say that it is difficult for any one index to prove to be a good benchmark.

In addition, the municipal bond market can have wide price swings between trading houses for the same bond. To correct that, for example, *The Bond Buyer* surveys five dealer institutions to obtain an average price for each of the issues in its Bond Buyer 40.

Where Is It Found?   The average is published in *Barron's* and *The Wall Street Journal* and also is available via America Online and Dow Jones News/Retrieval online database services.

How Is It Used for Investment Decisions?   The investor can use bond average and index information to determine the performance of bonds and bond mutual funds. The index indicates how well bonds are doing overall in the marketplace. A low index may indicate that poor returns are currently being earned and may signal a time to reallocate funds out of bonds and into a higher return alternative. On the other hand, a low index may indicate the time to buy bonds at a low price with the expectation of selling when prices are higher.

A conservative investor who wants to protect his principal from credit risk may buy U.S. guaranteed obligations. They are considered among the safest securities in the world, so their prices change largely based on varying interest rates. So rising indexes means falling yields on Treasuries, and vice versa.

The lone exception to this rule is mortgage-backed bonds. While principal is guarantee by government agencies, a "prepayment" risks exists. That is, investors may, and do, get repaid early when the underlying mortgages are refinanced or foreclosed upon. So when mortgage bond indexes are not performing as well as Treasury indexes, for example, an investor would know that refinancings or foreclosures are a concern.

Municipal bonds indexes measure price changes among these securities from state and local governments and their related agencies that are usually coveted for their tax-free income. This is an often underwatched area because many investors buy municipal bonds strictly for their yields and credit ratings.

Many corporate bonds provide a way for investors to earn extra income above Treasury issues without taking on major extra credit risk. The bond average gives the investor a clue as to how corporate bonds are performing in the marketplace on which he or she can base an intelligent investment decision. A strong market for corporate bonds may also be a good sign for the stock market, because bond prices often improve when the economy is good and the chances of default falls.

A corporate bond-investor willing to assume substantial risk may invest in junk bonds to achieve higher return. However, significant losses may occur. So a junk-bond investor should carefully monitor junk bond indexes.

Another risky way to boost returns from bonds is to own debt securities from foreign countries. Global bond indexes track the impact that price change and currency change have on these bonds for U.S. investors.

A Word of Caution: Serious bond investors compare performance and yields on various bond types to see which niche may be best for their portfolio. One way to do this is to compares yields earned on government bonds and corporate bonds of the same maturity. Further, a yield comparison may be made separately for short-term, intermediate-term, and long-term government and corporate bonds. (See Figure 12.)

The yield spread on similar maturity bonds is equal to the yield on corporate bonds minus the yield on government bonds. So, for example if, the yields on corporate bonds and government bonds are 7 and 6 percent, respectively, the yield spread is 1 percent.

DOW Jones 20 Bond Average

Utility Bonds	Coupon	Maturity
Commonwealth Edison	7.625%	6/01/2003
New York Telephone	7.375%	12/15/2011
Consolidated Natural Gas	7.25%	12/15/2015
BellSouth Telecom	7.625%	5/15/2035
BellSouth Telephone	6.375%	6/15/2004
Pacific Bell	7.125%	3/15/2026
Michigan Bell	7.0%	11/01/2012
Tucson Electric	7.65%	5/01/2003
Philadelphia Electric	7.375%	12/15/2001
Potomac Electric	7.0%	1/15/2018

Industrial Bonds	Coupon	Maturity
AT&T	7.125%	1/15/2002
du Pont	6.0%	12/01/2001
Bethlehem Steel	6.875%	3/01/1999
Atlantic Richfield	9.125%	3/01/2011
Sun Company	9.375%	6/01/2016
International Business Machines	6.375%	6/15/2000
Occidental Petroleum	10.125%	9/15/2009
Sears Roebuck	9.5%	6/01/1999
Inheres Busch Inc.	8.625%	12/01/2016
International Business Machines	7.0%	10/30/2025

FIGURE 12   (From www.dowjones.com, November 1998. With permission.)

Because government bonds have no default risk while corporate bonds do, a wider yield spread indicates traders' perception of a greater default risk with the corporate security. There will be a tendency to regress to narrow the range. That may mean that a currently attractive bond market niche may suddenly change when investors discover yield disparities. This can be true for any bond niche.

As an example, let us assume that the yield spread on similar maturity long-term government and corporate bonds is normally 2 percent. Yet for the current period, this spread has widened to 5 percent.

The wide spread in this example could indicate a buying opportunity for corporate bonds. It is likely that the gap will narrow in one of three ways.

1. Corporate rates will go down while government rates remain flat.
2. Corporate rates will go down while government rates will increase.
3. Corporate rates will remain flat while government rates rise.

As a result, the corporate bonds may outperform the government bonds on a total return basis. In this example, the investment strategy is to buy corporate bonds and sell government bonds. A rush of traders executing this strategy will dramatically narrow the opportunity to gain price profit from such a strategy.

Tool #11

Breadth (Advance–Decline) Index

What Is This Tool? The Breadth (Advance–Decline) Index, used in technical analysis, computes the net advances or declines in stocks on the NYSE for each trading day. A strong market exists when there are net advances while a weak market exists when there are net declines.

How Is It Computed?

$$\text{Breadth Index} = \frac{\text{Number of net advances or declines in securities}}{\text{Number of securities traded}}$$

Example: Assume net advancing issues are 230. Securities traded are 2145. The Breadth Index equals

$$\frac{\text{Net Advancing issues}}{\text{Number of securities traded}} = \frac{230}{2,145} = +0.107$$

The higher the positive percentage, the better because *more* stocks are increasing in price relative to those decreasing in price.

$$\text{Zweig Breadth Advance–Decline Indicator} = \frac{\text{10-day moving average of advancing issues}}{\text{10-day moving average of declining issues}}$$

Where Is It Found? The Breadth Index may be computed easily by referring to the financial pages of a newspaper. The market diary section of the paper will provide the number of advancing and declining issues along with the number of issues unchanged. The total number of issues traded equals the sum of these and is often provided as well. Some financial advisory publications calculate the index, relieving the investor from performing the computation. Martin Zweig's Breadth Advance–Decline Indicator is published in the *Zweig Forecast*. The financial news program *CNBC* reports it daily.

How Is It Used and Applied? Breadth analysis emphasizes change rather than level. The Breadth Index should be compared to popular market averages. Typically, there is consistency in their movement. In a bull market, an investor should be on guard against an extended disparity of the two. An example is when the Breadth Index gradually moves downward to new lows while the Standard & Poor's 500 Index reaches new highs.

TRADING DIARY

Supplied by Quotron, "QCHA" is the average percentage movement for all exchange listed stocks each day on an unweighted basis.

Market Advance/Decline Volumes

Daily	Aug. 21	22	23	24	25
NY Up	330,627	409,871	357,968	359,341	332,619
NY Off	323,694	334,786	444,552	374,225	282,615
% (QCHA)	−.06	+.08	−.03	+.08	+.14
Amex Up	21,508	9,577	31,613	27,238	7,837
Amex Off	5,416	19,548	7,289	5,377	17,621
% (QACH)	+.41	+.03	+.20	+.65	+.55
NASD Up	701,850	762,729	872,543	921,012	571,682
NASD Off	473,646	531,138	478,346	494,455	612,379
% (QCHAQ)	+.37	+.41	+.61	+.89	+.64

Market Advance/Decline Totals
Week ended last Friday compared to previous Friday

Weekly Comp.	NYSE	AMEX	Nasdaq
Total Issues	3,513	948	5,107
Advances	1,688	431	2,650
Declines	1,550	366	2,024
Unchanged	275	151	433
New Highs	254	65	214
New Lows	116	63	267

NYSE Common

Daily	Aug. 21	22	23	24	25
Issues Traded	2,776	2,784	2,779	2,776	2,772
Advances	1,073	1,160	1,044	1,169	1,184
Declines	1,277	1,160	1,306	1,185	1,140
Unchanged	426	464	429	422	448

NYSE Composite Daily Breadth

Daily	Aug. 21	22	23	24	25
Issues Traded	3,326	3,339	3,324	3,332	3,308
Advances	1,273	1,424	1,285	1,398	1,407
Declines	1,529	1,360	1,499	1,412	1,327
Unchanged	524	555	540	522	574
New Highs	85	91	98	69	75
New Lows	35	41	46	44	35
Blocks	15,107	17,801	19,018	17,904	13,967
Total (000)	901,884	1,001,893	1,050,315	1,000,795	826,774

AMEX Composite

Daily	Aug. 21	22	23	24	25
Issues Traded	782	760	752	777	768
Advances	297	271	292	319	303
Declines	304	300	286	281	269
Unchanged	181	189	174	177	196
New Highs	21	16	24	23	20
New Lows	11	23	24	18	9
Blocks	605	698	927	673	n.a.
Total (000)	42,514	47,322	56,849	49,276	41,652

Nasdaq

Daily	Aug. 21	22	23	24	25
Issues Traded	4,705	4,711	4,655	4,676	4,647
Advances	1,991	1,969	1,948	2,169	2,088
Declines	2,030	2,036	2,023	1,796	1,816
Unchanged	684	706	684	711	743
New Highs	69	69	66	76	82
New Lows	62	66	68	114	46
Blocks	13,637	15,771	16,448	18,394	n.a.
Total (000)	1,212,367	1,341,542	1,395,830	1,454,733	1,223,539

FIGURE 13A   Advance–Decline for Stocks. How newspapers cover advancing vs. declining stocks. (From *Barron's Market Week,* August 28, 2000, p. MW71. With permission.)

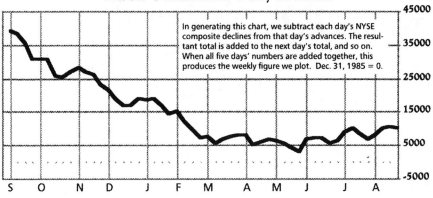

NYSE Cumulative Daily Breadth

In generating this chart, we subtract each day's NYSE composite declines from that day's advances. The resultant total is added to the next day's total, and so on. When all five days' numbers are added together, this produces the weekly figure we plot. Dec. 31, 1985 = 0.

FIGURE 13B   How a newspaper covers breadth.

A comparison also may be made of the Breadth Index over a 5- to 10-year period. The Breadth Index also may be compared to a base year or included in a 150-day moving average. (See Figure 13.)

How Is It Used for Investment Decisions?   The investor is interested in market direction to identify strength or weakness. Advances and declines usually follow in the same direction as standard market averages (e.g., Standard & Poor's 500 Index and the Dow Jones Industrial Average). However, they may go in the opposite direction at a market peak or bottom.

The investor can be confident of market strength when the Breadth Index and **a** standard market index are increasing. Securities may be bought because a bull market is indicated. If the indexes are decreasing, market weakness is indicated. Securities should not be bought in a bear market. In fact, securities held should be sold.

A Word of Caution: Historically, stock advances have exceeded declines. However, a sudden reversal can occur in the future and a net decline may, in fact, occur.

TOOL #12

BRITISH POUND

What Is This Tool? The currency of one of the United States' top allies and trading partners, the British pound, is one of the world's most important currencies. Its relationship to the U.S. dollar is a key to the global marketplace and is seen as a barometer of the United Kingdom's economic strength vs. the business climate in the United States.

How Is It Computed? It is typically quoted in newspapers and financial reports on television and radio in terms of its relationship to the U.S. dollar. As of December 1998, the British Pound was not to be included in the new "euro" currency that will unite major currencies across Western Europe. If the pound is at 1.4, that means each U.S. dollar buys 0.71 pounds. To figure out what 1 pound equals in U.S. currency, this formula is used.

Example: If $1 buys 0.71 pounds, then 1 pound is equal to $1 divided by 1.4 or $1.40.

Where Is It Found? Currency rates are listed daily in most major metropolitan newspapers as well as national publications such as *The New York Times, The Wall Street Journal,* and *USA Today* and on computer services such as America Online or at web sites on the Internet such as www.bloomberg.com or quote.Yahoo.com.

How Is It Used for Investment Decisions? For American investors buying British securities, the pound's movement is a key part of the profit potential in the investment. If the pound rises after an investment in British securities is made, the value of those stocks, bonds, or property to a U.S. investor will get a boost from the currency. That is because when the investment is sold, the stronger pound will generate more dollars when the proceeds are converted to the U.S. currency. Conversely, a weak pound will be a negative to a British investment for a U.S. investor. In some cases, the movement of the pound can also be viewed as an indicator of British economic health. A strong pound can signal a buoyant economy, a possible indication to buy British stocks. The pound's strength, however, should be verified not only against the U.S. dollar but against other major currencies. (See Figure 14.)

A Word of Caution: The pound's strength vs. the dollar can be distorted by prevailing interest rates in each country. For example, a movement by the British central bank to slow down the economy by boosting British interest rates—two potential negatives for stock prices—might also increase the pound's price vs. the dollar if U.S. rates are stagnant or falling. This scenario of a rising pound might give an incorrect reading on the potential of buying British stocks.

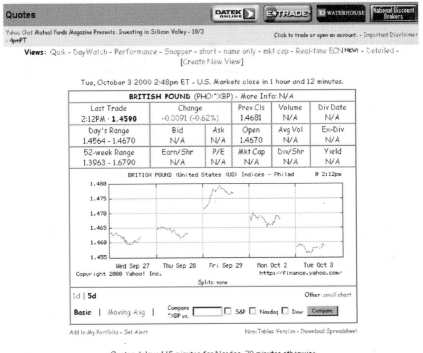

FIGURE 14 How a web site watches the British pound. (Reproduced with permission of Yahoo! Inc.© 2000 by Yahoo! Inc. YAHOO! and the YAHOO! Logo are trademarks of Yahoo! Inc.)

Also, widely quoted currency rates are typically for transactions of $1 million or more. Consumers attempting to use such figures to determine currency rates for foreign travel should expect to get somewhat less favorable exchange rates. (See Figure 15.)

Also See: Euro, German Deutsche Mark, Japanese Yen.

Currency vs. Returns

Index	3-Year Total Return in U.S. dollars	3-Year Total Return in Local Currencies	Currency Impact to U.S. Investor
EAFE	3.8%	7.4%	−3.7%
Europe	18.8%	19.5%	−0.6%
Japan	−17.5%	−8.7%	−8.8%
Pacific (without Japan)	−13.5%	−9.8%	−3.7%
World (without the United States)	3.9%	7.5%	−3.6%
World	12.2%	13.7%	−1.5%

FIGURE 15   Currency Impact on Investing. Here is a look at how changes in currency rates can alter investment returns for U.S. investors. This chart tracts total return in key global stock indexes in both U.S. dollars and local currencies. Indexes are by Morgan Stanley Capital International and the data period is the three years ending on September 30, 1998. (From Morgan Stanley Capital International. With permission.)

Tool #13

Bullishness Indicators

What Are These Tools?   They are numerous surveys of investors seeking to determine whether key market watchers and players are likely to be buyers (bulls) or sellers (bears) of stocks in the near term.

The Bridge Market Barometer is a sentiment indicator developed by canvassing a regular set of market participants and asking them if they are bullish or bearish on a given market one week from today. Sentiment readings typically have been used as a contrarian indicator, although in recent years many markets have shown that very high readings are actually leading bullish indicators.

How Are They Computed?   Various groups conduct polls of various sets of investors, ranging from professional traders to the small stockholder. Typically, they ask the survey sample one key question: "Are you bullish/bearish/neutral on the stock market?" The survey then adds up the percentage of those polled and their responses. A "bullish" reading can be anything over 50 percent or, after a prolonged market drop, a turn from increasing bearishness to increasing bullishness.

Where Are They Found?   *Barron's* publishes four such surveys each week: a poll of investment advisers and/or traders by Investors Intelligence of New Rochelle, NY, Consensus Inc. of Kansas City, MO, and Market Vane, plus a poll of small investors by the American Association of Individual Investors of Chicago.

The Bridge Market Barometer is calculated by polling a regular set of market participants and asking them if they are bullish or bearish on a given week. In general, a reading above 50.0 would indicate bullish psychology within a broad spectrum of market participants. The Bridge Market Barometer weights the level of bullish or bearish expectations from respondents, so the final barometer reading often will be slightly different from the absolute number of bullish responses. The Bridge Barometer is updated each Thursday and available on http://www.bridge.com/front/. (See Figure 16.)

How Are They Used for Investment Decisions?   Such polls can be used to measure the market trend's true strength, often in a contrarian way. In a rising market or one near its peak, growing bearish feelings may indicate the market may be headed higher because healthy skepticism abounds. A hot market with strong bullish feelings could be a signal or overheated investor expectation, howerver, that might be a sell signal. In a falling market or one near a possible bottom, the slightest turn to bullishness could be a buying signal. At such a low point for stocks, watch for any hint that bearish sentiment is ebbing.

How the Bridge Barometer Fluctuates

Date	Percent Bridge Barometer	Bullish Response	Simple Figure Change
11/5/98	62.9	67	27
10/29/98	49.1	50	−2
10/22/98	58.6	65	17
10/15/98	53.2	59	7
10/8/98	39.3	44	−17
10/1/98	46.5	45	−7
9/24/98	50.9	52	2
9/17/98	56.8	64	15
9/10/98	38	25	−25
9/3/98	40.1	29	−17
8/27/98	46.5	55	−7
8/20/98	56.2	52	13
8/13/98	55.9	59	13
8/6/98	52.9	62	6
7/30/98	74.7	93	37
7/23/98	60.8	67	26
7/16/98	57.7	64	17
7/9/98	60	58	19
7/2/98	72.6	79	43

FIGURE 16   A look at the Bridge® Barometer during the turbulent summer and fall of 1998. (From Bridge Information Systems, Inc. With permission.)

A Word of Caution: Such surveys often give conflicting signals, leading to tricky analysis. Do not tend to lean toward the surveys of professional traders in such cases. Remember that many studies show that Mom and Pop Investors often do quite well on their own.

Tool #14

Cash Investments

CERTIFICATE OF DEPOSIT YIELDS

What Are These Tools?   Certificates of deposits are accounts offered by banks that pay set interest rates for a stated amount of time. Yields on certificates of deposit provide a gauge of what bankers are paying on short-term, fixed-maturity savings accounts. CDs are popular investments for both small and deep-pocketed savers to stash their cash holdings.

How Are They Computed?   Several different organizations track bank deposit rates. Bank Rate Monitor and 100 Highest Yields of Florida, Bradshaw Financial of Massachusetts, and Banxquote of New York poll banks and savings institutions about rates offered to consumers across the nation. They publish weekly reports on how savings rates have changed, both nationwide and in major U.S. cities.

The Federal Reserve also compiles weekly averages of so-called jumbo CDs, which are deposits of $100,000 and more, placed mainly by institutional investors. Various electronic services report daily on rates paid on these savings accounts.

Where Are They Found?   Many daily newspapers (both national and local) carry information on CD rates, compiling lists of offerings from local banks as well as national high-rate leaders. Banks are aggressive newspaper advertisers so a keen shopper can find a good offer in various parts of a newspaper. The Internet is also a good source. Try www.banx.com for Banxquote's Web site or imoneynet.com for Bradshaw's Web site. (See Figure 17.)

How Are They Used for Investment Decisions?   Certificates of deposits are popular savings tools for the small investor. When these rates are high or rising, such investors may flock to CDs and unlikely to be actively buying stocks or bonds. That can be a negative signal to those markets. Conversely, when CD rates are low or falling, small savers tend to look to markets like stocks, real estate, or bonds to improve their returns.

CD rates also can signal the banking industry's outlook for the economy. The movement of CD rates, particularly when compared to broader market interest rates, can suggest how bankers expect interest rates to move. The aggressiveness of the bankers' moves can show a willingness to lend if they are pushing up CD rates or the lack of good loans may be signaled if CD rates are plunging.

A Word of Caution: CDs are relatively non-liquid unless a saver wants to pay early withdrawal fees. That expense can dramatically affect the yield on a CD investment. Also, many attractively yielding CDs come with higher minimum deposit requirements.

Savers' scoreboard

Highest CD yields this week

6-month

Bank, phone	Yield
DeepGreen Bank U, 888-576-9238	7.15%
VirtualBank U, 877-998-2265	7.10%
giantbank.com, 877-446-4200	7.05%
1-year	
DeepGreen Bank U, 888-576-9238	7.30%
VirtualBank U, 877-998-2265	7.25%
AFBA Industrial Bk, 800-776-2265	7.20%
2½-year	
USAccess Bk, 877-369-2265	7.30%
Provident Bank, 800-335-2220	7.29%
Providian Bank, 800-414-9693	7.27%
5-year	
Provident Bank, 800-335-2220	7.57%
Key Bank USA, 800-872-5553	7.46%
Eastern Svgs Bk, 800-787-2265	7.40%

Source: Bankrate.com, North Palm Beach, Fla., www.bankrate.com

Money market funds

The biggest money market mutual funds open to individuals and their seven-day annualized yields.

Fund (ranked by size)	This week	Last week	6 mos. ago
Smith Barney Cash Port	6.10%	6.10%	5.32%
Vanguard Prime MMF/Retail	6.32%	6.33%	5.62%
Merrill Lynch CMA Money	6.10%	6.12%	5.44%
Fidelity Cash Reserves	6.23%	6.24%	5.69%
Schwab Money Market	5.95%	5.95%	5.27%
Average taxable money fund	**6.00%**	**6.01%**	**5.28%**

Source: Money Fund Report, 800-343-5413; www.imoneynet.com

FIGURE 17 CD Rates. How a newspaper tracks certificate of deposit yields. (Copyright 2000, USA TODAY. Reprinted with permission.)

GUARANTEED INVESTMENT CONTRACT YIELDS

What Are These Tools? Guaranteed investment contracts (GICs) are a popular fixed income investment offered by insurance companies. Most investors see them in 401(k) retirement plans offered at the workplace. The contracts act much like bankers' certificates of deposit, promising a fixed rate of return over a specified time. In employer retirement savings' plans that offer GICs to workers, GICs on average take in more than one-third of the money in the plan.

How Are They Computed? Average GIC yields are compiled by several firms such as T. Rowe Price of Baltimore that survey major insurance companies to see what yields and maturities they are offering. GICs run in excess of $1 million.

Where Are They Found? The Rowe Price report on GICs appears daily in *The Wall Street Journal* while Fiduciary Capital's review is published weekly in *Barron's*. (See Figure 18.)

GUARANTEED INVESTMENT CONTRACTS

[1]Current Treasury quotes, as reported in the Wall Street Journal. Total issuers quoting: 37. [2]Spreads between GIC/BIC high and Treasury with the same maturity. All rates for non benefit responsive, investment only contracts, net of expenses, no commissions. Source: Fiduciary Capital Management Inc., Woodbury, Conn.

Week Ending: August 25, 2000

Rates	Three Year		Five Year		Seven Year	
	Comp.	Simple	Comp.	Simple	Comp.	Simple
$1 Million						
High	7.60	7.57	7.64	7.62	7.66	7.66
Average	7.21	7.15	7.28	7.22	7.29	7.21
Low	6.52	6.54	6.56	6.59	6.54	6.60
# Issuers	14	16	14	16	12	12
$5 Million						
High	7.65	7.62	7.69	7.67	7.71	7.71
Low	6.75	6.60	6.67	6.60	6.67	6.60
# Issuers	20	24	20	24	17	18
High	7.65	7.62	7.69	7.67	7.71	7.71
$10 Million						
Average	7.19	7.15	7.26	7.20	7.29	7.20
Low	6.81	6.60	6.71	6.60	6.76	6.60
# Issuers	18	22	18	22	16	17
High	7.56	7.53	7.59	7.51	7.62	7.50
$25 Million						
Average	7.18	7.09	7.25	7.16	7.30	7.18
Low	6.91	6.60	6.92	6.60	6.92	6.60
# Issuers	12	13	12	13	11	12
		6.13		6.01		5.89
Treasuries[1]						
$1 Million	1.47	1.44	1.63	1.61	1.77	1.77
Treasuries-GIC/BIC Spreads[2]						
$10 Million	1.52	1.49	1.68	1.66	1.82	1.82
$25 Million	1.43	1.40	1.58	1.50	1.73	1.61
Average	1.49	1.46	1.64	1.61	1.79	1.76

FIGURE 18 Guaranteed Investment Contracts. How a newspaper covers guaranteed investment contracts' yields. (From *Barron's Market Week,* August 28, 2000, p. MW76. With permission.)

How Are They Used for Investment Decisions? An investor can review GIC yields to see how his offerings in an employer's retirement plan compare to other fixed-income choices both in the plan and ones that are offered elsewhere. Changes in GIC yields also can be used as an indicator of how insurance company executives see the outlook for interest rates much like certificate of deposit rates signal bankers' intentions.

If, for example, GIC rates are falling faster than other similar rates, that can be viewed as a sign that insurers do not actively want money now because they see an upcoming steep drop in rates. When rates do go lower, insurance companies can profit by then scooping up large inflows at those cheaper rates.

Conversely, if GIC rates are rising faster than competing rates, that can be viewed as a bet by insurers that they want to draw in large sums of money quickly because they think rates may eventually go much higher. When rates do go higher, insurance companies will have profited by locking large inflows at those lower rates.

A Word of Caution: Although GICs look and sound like bank products, they do not have the same safety. The only guarantee backing GIC investments is the promise of the insurance company or companies underwriting the contract. A bank account, by comparison, is insured by the U.S. government up to $100,000. Because of this risk, many employers now offer pools of GICs to their employees to spread the risk of default among several insurers.

MONEY MARKET FUND AVERAGE MATURITY

What Is This Tool? This is information reported weekly on the maturity of securities held by money market mutual funds.

How Is It Computed? The figures reflect the average of all securities held by money funds and is tracked weekly by IBC Financial of Massachusetts. Money funds are restricted to securities of no more than one year in maturity and cannot have an average maturity more than 90 days. Many newspapers such as *The Orange County Register* and the *Boston Globe* run both fund and industrywide figures weekly as does *Barron's*. The information on individual funds also can be obtained from the fund's management.

How Is It Used for Investment Decisions? Changes in the average maturity can be seen as an indication of where shortest-term interest rates are headed.

When managers believe that rates are headed lower, they extend their maturities to lock in what they believe are currently higher rates. When fund managers think rates are headed higher, they shorten the average maturity so they can have the most money available to buy the expected higher-yielding paper in the near future. The IBC industrywide average serves as a consensus for the overall leanings of all fund managers.

When comparing individual funds, relative maturities can show which fund or funds will better respond to such rate moves. Longer maturity funds should do best if rates fall. Shorter maturity funds excel when rates rise. Of course, with a maximum maturity of 90 days, such advantages will diminish quickly.

Also See: Cash Investments: Money Market Fund Yields.

MONEY MARKET FUND YIELDS

What Are These Tools? These are a measurement of the return on money market mutual funds that are popular alternatives to bank accounts and other cash investments. Money funds are managed to maintain a stable $1 per share price.

How Are They Computed? The money fund yield is typically viewed in three ways: a 7-day yield compounded over a year, a 30-day yield compounded over a year, and a 12-month total return.

The 7-day and 30-day yields reflect dividends the money funds have collected in a recent week and month, respectively, less costs such as management fees. Those returns are then annualized, that is, assumed to be collected at the same rate for a year.

The annual return figure reflects both dividends collected as well as any capital gain earned minus any expenses. Accounting rules allow those capital gains to be shown to investors as yield enhancements rather than per share price changes.

Where Are They Found?   IBC's 7-day yields are published weekly in many daily newspapers such as *Dallas Morning News, The New York Times,* and *The Orange County Register.* The 30-day yield and total return figures are typically what is quoted by mutual fund companies in sales literature. Internet users can also check IBC's Web page at imoneynet.com. (See Figure 19.)

How Are They Used for Investment Decisions?   IBC's average of all money fund yields are used as both an indicator of what very short-term investments are returning and a barometer to measure all funds' performance. An investor must be careful, however, when comparing yields of funds.

Although these funds are managed very conservatively, there are some subtle safety differences that should be reviewed. Some funds own a mixture of corporate, banking, and government obligations. Others take a slightly less risky path, holding only U.S.-backed issues. Such funds tend to pay out less.

A Word of Caution: Many fund management companies temporarily waive money fund fees (which can cut as much as a full percentage off the yields) as an inducement to invest. These discounts can be misleading, so investors are advised to check out a high-paying fund carefully. An investor should query whether the fund is offering such a discount, how long the discount will last, and what the fund would have yielded if the full fees were charged.

THE LARGEST MONEY MARKET FUNDS

Here are the 15 largest taxable money market funds and their compounded 30-day yields as of September 12.

Fund	Now	1 wk. ago	6 mo ago	(800) Phone	Min. Initial Deposit	Total Assets, Billions
Smith Barney Cash 'A'	6.10	6.10	5.32	544-7835	1,000	52.81
Vanguard Prime	6.32	6.33	5.62	662-7447	3,000	45.00
Merrill Lynch CMA Money	6.10	6.12	5.44	221-7210	5,000	44.00
Fidelity Cash Reserves	6.23	6.24	5.69	544-8888	2,500	43.19
Schwab Money Market	5.95	5.95	5.27	266-5623	1,000	40.05
Schwab Value Advantage	6.30	6.29	5.63	266-5623	25,000	34.08
MS Dean Witter Active Assets	6.15	6.18	5.56	869-3326	10,000	23.72
MorgStan Dean Wit Lq. Ast.	6.09	6.09	5.50	869-3863	0	19.95
PaineWebber RMA	6.15	6.14	5.36	762-1000	0	19.92
Centennial Money Market	6.07	6.02	5.33	525-9310	500	19.67
Prudential Command Money	6.20	6.21	5.52	222-4321	10,000	14.38
Wells Fargo MMF	5.86	5.86	5.22	956-4442	2,500	13.81
Alliance Capital Reserves	5.68	5.71	5.00	221-5672	1,000	11.48
Merrill Lynch Retire	6.14	6.15	5.61	221-7210	0	10.23
Evergreen MMF	5.57	5.58	0.00	544-8888	20,000	10.14

Source: iMoneyNet, Ashland, Mass., 508-616-6100, **www.imoneynet.com**

FIGURE 19   How a newspaper tracks money market mutual finds. (From *Miami Herald,* September 18, 2000, p. 46. With permission.)

Also, compounded yields assume that short-term market conditions will continue. That can be misleading at times. One can also track "straight" money funds yields (that is, without compounding) to get a, perhaps, more sanguine view of the money fund world.

Also See: Cash Investments: Money Market Fund Average Maturity, Interest Rates: Discount, Fed Funds, and Prime Interest Rates: Three-Month Treasury Bills.

Tool #15

Changes: Stocks Up and Down

What Is This Tool? This is a listing (usually daily) of the stocks having the highest percentage increase in price and the highest percentage decrease in price for the trading period. The stocks are listed in the order of their percentage change. Typically, 10 advancing stocks and 10 declining stocks are listed. There is a separate listing for the NYSE, AMEX, and NASDAQ issues.

How Is It Computed? A tabulation is made of the stocks on each exchange in terms of their percentage increase or decrease in market price. The tables typically use only stock prices above $1.

Where Is It Found? The up changes and down changes appear in the financial pages of newspapers and magazines such as *Barron's, The New York Times,* and *The Wall Street Journal.* Figure 20 is a sample listing of NYSE, AMEX, and NASDAQ Stocks with the largest percentage changes up and down.

How Is It Used for Investment Decisions? In looking at price increases or decreases, consideration should be given to the trading volume. For example, a price increase on heavy volume is a stronger indicator of movement than a price increase on low volume. If a stock appears frequently under both the up changes and down changes listings, it is a volatile issue because it fluctuates in price drastically. This is typically associated with a speculative issue that implies risk.

The investor must determine whether a stock with a significant price increase can sustain its upward momentum. If it can, the stock might be attractive. On the other hand, if the sharp increase in price cannot be sustained and is temporary, the market price of the stock might fall because it is overvalued. In such a case, the investor should not buy the stock.

If there is a sharp decline in the market price of a stock, the investor should ascertain whether the issue is in a downward trend or whether it is now undervalued and should be bought. Further, a sharp move in stock price may be owing to specific good or bad news.

A Word of Caution: Also, a company experiencing a significant percentage change in price in a given day may be vulnerable to increased price volatility the next day. One scenario is that the stock was oversold or overbought during that volatile day and that other investors may viewing it as undervalued or overvalued and begin reversing the previous day's trend.

Conversely, when a stock is a large percentage loser, some traders may think the worst of the news is not out and continue to apply selling pressure. When a stock is a

NYSE winners

Name	Last	Chg	Pct
DotHill	8.88	+1.56	+21.4
Omnova n	6.00	+1.00	+20.0
BrillChA s	34.00	+4.88	+16.7
Dave&B	8.00	+1.13	+16.4
GC Cos	14.88	+1.88	+14.4
Entrade	5.00	+.63	+14.3
Telemig	65.00	+8.00	+14.0
TB Woods	11.25	+1.38	+13.9
Lexmark	67.50	+8.00	+13.4
CmtyHlt n	23.75	+2.69	+12.8
WallCS	11.63	+1.31	+12.7
THiltgr	10.88	+1.19	+12.3
Oceaner	17.44	+1.88	+12.0
PrisonR	2.38	+.25	+11.8
ChmpE	6.63	+.69	+11.6

NYSE losers

Name	Last	Chg	Pct
BestBuy	61.88	-9.13	-12.9
Gap	22.38	-3.19	-12.5
CKE Rst	4.00	-.50	-11.1
Target s	23.19	-2.88	-11.0
Matlack	2.13	-.25	-10.5
Consc pfV	11.64	-1.36	-10.5
PrecDr g	34.00	-3.56	-9.5
BritSky	95.50	-9.50	-9.0
Tndycft	2.06	-.19	-8.3
Neff Cp	4.88	-.44	-8.2
AberFitc	23.19	-2.06	-8.2
Heico	15.50	-1.38	-8.1
IT Gp	4.38	-.38	-7.9
Acptins	5.88	-.50	-7.8
Penney	14.00	-1.19	-7.8

FIGURE 20　How a newspaper tracks percentage winners and losers for stocks. (From *Orange County Register* (CA), September 1, 2000, p. B6. With permission.)

large percentage gainer, some traders may see it on an up and down list and think things will only get better for this company and add further buying pressure to the issue.

Tool #16

Charting

What Is This Tool?   Charts are used to evaluate market conditions including the price and volume behavior of the overall stock market and of individual securities.

How Is It Prepared?   The basic types of charts are line, "high–low–close" bar charts, and point-and-figure. On line charts (see Figure 21) and bar charts (see Figure 22), the vertical axis shows price and the horizontal axis shows time.

On a line chart, ending prices are connected by straight lines. On a bar chart, vertical lines appear at each time period, and the top and bottom of each bar shows the high and low prices. A horizontal line across the bar marks the ending prices.

Point-and-figure charts (see Figure 23) show emerging price patterns in the market, in general, and for specific stocks. Typically, only the ending prices are charted. An increase in price is denoted by an "X" while a decrease in price is shown as an "O."

There is no time dimension in a point-and-figure. A column of Xs shows an upward price trend while a column of Os reveals a downward price trend.

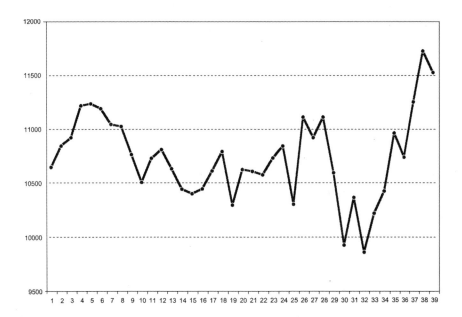

FIGURE 21   How a line chart of the Dow Jones Industrial Average stock looks.

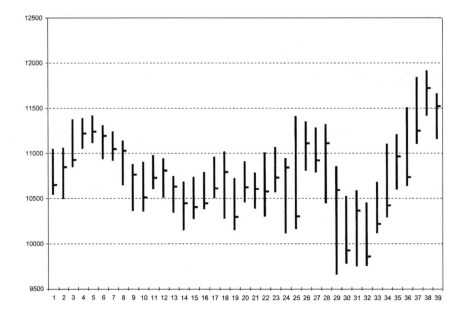

FIGURE 22   How a bar chart of the Dow Jones Industrial Average looks.

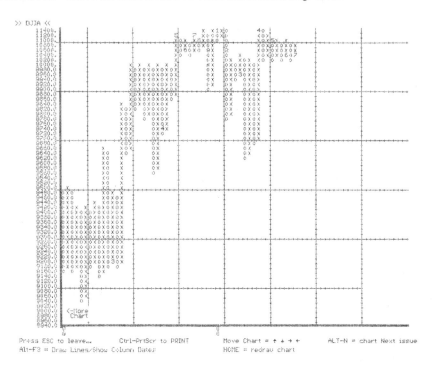

FIGURE 23   Point-and-Figure Chart. How a point and figure chart of the Dow Jones Industrial Average looks.

In point-and-figure charts, there is a vertical price scale. Plots on the chart are made when a price changes by a predetermined amount.

Significant price changes and their reversal are depicted. What is significant is up to the individual investor. The investor can use either ending prices or inter day prices, depending on time constraints. The usual predetermined figures are 1 or 2 points for medium-priced stocks, 3 or 5 points for high-priced stocks, and 1/2 point for low-priced stocks. Most charts contain specific volume information.

The investor should plot prices representing a trend in a single column, moving to the next column only when the trend is reversed. The investor will usually round a price to the nearest dollar and start by plotting a beginning rounded price. Nothing new appears on the chart if the rounded price does not change. If a different rounded price occurs, the investor plots it. If new prices continue in the same direction, they will appear in the same column. A new column begins when there is a reversal.

Where Is It Found?   Various market-watch services, financial magazines and newspapers, and brokerage research reports provide charts as well as various web sites such as www.bigcharts.com or www.bloomberg.com on the Internet.

How Is It Used and Applied?   A chart pattern may be studied to predict future stock prices and volume activity. Charts may cover historical data of one year or less for active stocks or several years for inactive stocks.

Point-and-figure charts provide data about resistance levels (points). Breakouts from resistance levels indicate market direction. The longer the sideways movement before a break, the more the stock can increase in price.

How Is It Used for Investment Decisions?   The investor may use charts to analyze formations and spot buy and sell indicators.

The investor can use these charts to determine whether the market is in a major upturn or downturn, and whether the trend will reverse. The investor also can see what price may be accomplished by a given stock or market average. Further, these charts can help the investor predict the magnitude of a price swing.

A Word of Caution: Historical trends in prices may not result in future price trends because of changing circumstances in the current environment.

Also See: Japanese Candlestick Charts

TOOL #17

COMMODITIES INDEXES

CRB INDEXES

What Are These Tools? Bridge/CRB, formerly Commodity Research Bureau (CRB), has two widely watched benchmarks for commodity prices: the CRB Bridge Spot Price Index and the CRB Bridge Futures Price Index.

The CRB Spot Price Index is based on prices of 23 different commodities, representing livestock and products, fats and oils, metals, and textiles and fibers, and it serves as an indicator of inflation.

The CRB Futures Price Index is the composite index of futures prices that tracks the volatile behavior of commodity prices. As the best-known commodity index, the CRB index, produced by Bridge Information Systems, was designed to monitor broad changes in the commodity markets. The CRB index consists of 21 commodities.

In addition to the CRB Futures Index, nine subindexes are maintained for baskets of commodities representing currencies, energy, interest rates, imported commodities, industrial commodities, grains, oil-seeds, livestock and meats, and precious metals. All indexes have a base level of 100 as of 1967, except the currencies, energy, and interest rates indexes, which were set at 100 as of 1977.

How Are They Computed? The CRB Futures Index can be thought of as a three-dimensional index. In addition to averaging the prices of all 21 components, the index also incorporates an average of prices over time for each commodity. The price for each commodity is the simple average of the futures prices for a nine-month period.

Example: The average price for wheat contracts traded on the Chicago Mercantile Exchange in July 1999 would be determined as the average for the following five contract months: August, October, and December, 1999, and February and April, 2000. Mathematically,

$$\text{Wheat average} = \frac{\text{Aug. 1999} + \text{Oct. 1999} + \text{Dec. 1999} + \text{Feb. 2000} + \text{Apr. 2000}}{5}$$

For other commodities, there may be more or less than five contract months in the average.

The average prices of the 21 component commodities are then geometrically averaged and the result is divided by 53.0615 (the 1967 base-year average for these commodities); multiplied by 0.95035 and by 100. Mathematically,

$$\text{CRB Futures Index} = \frac{\text{Cattle avg.} \times \cdots \times \text{Wheat avg.} \times 0.95035 \times 100}{53.0615}$$

The factor 0.95035 amounts for an adjustment necessitated by the index's changeover (July 20, 1987) from 26 commodities averaged over 12 months to 21 commodities averaged over 9 months. The index is multiplied by 100 in order to convert its level into percentage terms. In other words, the CRB Futures Index involves both geometric and arithmetic averaging techniques.

Where Are They Found? *The Wall Street Journal* tracks this index as do some other daily newspapers. Internet users can check Bridge's Web site at www.crbindex.com for updates on this index. The CRB chart frequently appears in the cable network *CNBC*. (See Figure 24.)

How Are They Used for Investment Decisions? The CRB Futures Index serves as the basis for cash-settled futures contracts that are traded on the New York Futures Exchange (NYFE) under the commodity code CR. Higher commodity prices, for example, can signal inflation, which in turn can lead to higher interest rates and yields and lower bond prices. Higher interest rates tend to depress the stock market as well. Conversely, lower commodity price can signal deflation and economic weakness and push interest rates lower.

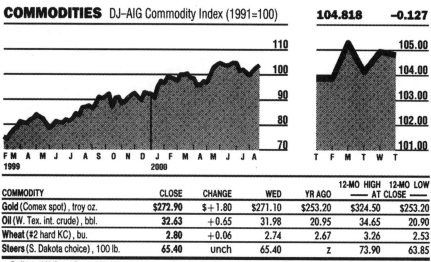

COMMODITIES DJ–AIG Commodity Index (1991=100)   **104.818   –0.127**

COMMODITY	CLOSE	CHANGE	WED	YR AGO	12-MO HIGH	12-MO LOW
					——— AT CLOSE ———	
Gold (Comex spot), troy oz.	$272.90	$+1.80	$271.10	$253.20	$324.50	$253.20
Oil (W. Tex. int. crude), bbl.	32.63	+0.65	31.98	20.95	34.65	20.90
Wheat (#2 hard KC), bu.	2.80	+0.06	2.74	2.67	3.26	2.53
Steers (S. Dakota choice), 100 lb.	65.40	unch	65.40	z	73.90	63.85

a- Reflects 6% Bond Buyer Municipal Index, from Dec. 22,1999.

FIGURE 24 CRB Index. Here is how a newspapers tracks commodity prices. (From *The Wall Street Journal,* August 25, 2000, p. C1. With permission.)

A Word of Caution: In order to gauge inflation pressure from raw goods, you should also look at popular price indexes such as the Producer Price Index (PPI).

Also See: Economic Indicators: Inflation, Economic Indicators: Interest Rates.

THE ECONOMIST COMMODITIES INDEX

What Is This Tool?   The indicators gauge commodity spot prices and their movements.

How Is It Computed?   The index is a geometric weighted-average based in international trade on the significance of spot prices of major commodities. The index is designed to measure inflation pressure in the world's industrial powers. It includes only commodities that freely trade in open markets, eliminating items like iron and rice. Also, this index does not track precious metal or oil prices. The commodities tracked are weighted by their export volume to developed economies.

Where Is It Found?   The index information may be obtained from Reuters News Services or in *The Economist* magazine.

How Is It Used for Investment Decisions?   The index may be used as a reflection of worldwide commodities prices, enabling the investor to determine the attractiveness of specific commodities. The investor may enter into futures contracts. The indicators also may serve as barometers of global inflation and global interest rates.

Tool #18

Contrarian Investing

CONTRARY OPINION RULE

What Is This Tool? Contrary opinion is a sentiment indicator in which, after finding out what most investors are doing, the investor does the opposite. The rationale is that popular opinion is usually wrong. The rule presumes that the crowd is typically incorrect at major market turning points.

Where Is It Found? The majority opinion is normally reflected in the mainstream news media including reports and articles in newspapers, on television, and in popular magazines such as *Newsweek* and *Time.* Other sources include financial publications, business books for lay people, and financial newsletters.

How Is It Used and Applied? A careful reading of the popular press is required. When one sees published and televised news detailing market activity, this logic says that this news usually has already been reflected in the price of securities. Therefore, the publicizing of the news is probably the end (instead of the beginning) of a move.

How Is It Used for Investment Decisions? An investor following a contrarian strategy does the opposite of what most investors are doing. If everyone is pessimistic, the investor concludes that it is probably the time to buy. If everyone is optimistic, the investor believes that it is probably the time to sell. The investors should compare the news stories with other technical and fundamental indicators.

The investor may find good buys for company stocks that are out of favor because of an oversold situation. However, these stocks should possess fundamental values based on the company's financial condition.

A Word of Caution: There may be instances in which what most investors are doing is the right strategy. The investing public may be buying securities during what is, in fact, a bull market. This is exactly what occurred during much of the 1990s, much to the dismay of many bearish Wall Street analysts.

Also See: Trading Volume Gauges: Odd-Lot Theory.

INDEX OF BEARISH SENTIMENT

What Is This Tool? The index is based on a reversal of the recommendations of investment advisory services as contained in their market letters. Such services are considered to be proxy for "majority" opinion.

This index operates according to the contrary opinion rule: Whatever the investment advisory services recommend, the investor should do the opposite. It is a technical investment analysis tool. *Investors Intelligence* believes that advisory services are trend followers rather than anticipators. They recommend equities at market bottoms and offer selling advice at market tops.

How Is It Computed?

$$\text{Index} = \frac{\text{Bearish investment advisory services}}{\text{Total investment advisory services}}$$

Investors Intelligence believes that when 42 percent or more of the advisory services are bearish, the market will go up. On the other hand, when 17 percent or less of the services are bearish, the market will go down.

Where Is It Found?   The Index of Bearish Sentiment is published by *Investors Intelligence* (914-632-0422). It can be found in *Barron's*. The index was originally developed by A.W. Cohen of Chartcraft.

Example: Of 200 investment advisory services, 90 of them are bearish on the stock market. The Index equals 0.45 (90/200).

Because 45 percent of the advisory services are pessimistic about the prospects for stock, or more than the 42 percent benchmark, the investor should buy securities.

How Is It Used for Investment Decisions?   A movement toward 10 percent means that the Dow Jones Industrial Average is about to go from bullish to bearish. When the index approaches 60 percent, the Dow Jones Industrial Average is headed from bearish to bullish.

The investor should use this index in predicting the future direction of the securities market based on contrary opinion. If bearish sentiment exists, a bull market is expected, and the investor should buy stock. If bullish sentiment exists, a bear market is likely, and the investor may consider selling securities owned.

A Word of Caution: Other measures of stock performance should be used in conjunction with this index. (See Figure 25.)

INVESTOR SENTIMENT READINGS

In Investors Intelligence's poll, the correction figure represents advisers who are basically bullish, but are looking for some sort of short-term weakness. High bullish readings in that poll, in Consensus Inc., or in Market Vane's usually are signs of market tops; low ones, market bottoms.

Investors Intelligence

	Last Week	Two Weeks Ago	Three Weeks Ago
Bulls	43.5%	47.1%	48.6%
Bears	33.7	31.7	33.6
Correction	22.8	21.2	17.87

Source: Investors Intelligence, 30 Church Street, New Rochelle, N.Y. 10801 (914) 632-0422.

Consensus Index

Bullish Opinion	62%	53%	47%

Source: Consensus Inc., 1735 McGee Street, Kansas City, Mo. 64108 (816) 471-3862.

AAII Index

Bullish	57.7%	59.4%	54.1%
Bearish	23.1	18.8	11.5
Neutral	19.2	21.9	34.4

Source: American Association of Individual Investors, 625 N. Michigan Ave., Chicago, Ill 60611 (312) 280-0170.

Market Vane

Bullish Consensus	38%	32%	33%

Source: Market Vane, P.O. Box 90490, Pasadena, CA 91109 (626) 395-7436.

FIGURE 25   Contrarian Indicators. How a newspaper tracks investor sentiment readings. (From *Barron's Market Week,* August 25, 2000, p. MW77. With permission.)

Tool #19

Credit Ratings

What Are These Tools?   Bonds and preferred stock issues are rated for credit-worthiness by various agencies for investment safety.

The major agencies are Standard & Poor's of New York, Moody's of New York, Fitch IBCA of New York, and Duff & Phelps of Chicago. These firms' analysts weigh the probability that interest payments will continue and the chances for repayment of principal.

Ratings help investors evaluate risks and aid the market in setting prices for various securities. Higher-rated securities attract more investors and, thus, trade at higher prices with lower yields. Low-rated paper, in turn, comes with lower prices and loftier yields.

How Are They Computed?   Each rating agency has different criteria and scales to weigh the credit risks of various securities. Bond and preferred stock issues are rated at initial offerings and then reviewed on a regular basis and after major economic events hit the issuer. Many rating systems' reviews, resulting in the bond rating, parallel grade-school report cards. That means As are the best and Cs and Ds are the worst.

All ratings assume that securities of the U.S. Treasury, in other words those that are backed by the full faith and credit of the U.S. government, warrant the highest rating for credit safety.

Where Are They Found?   All the agencies publish their ratings on a regular basis and when major changes are made. There are often news accounts in daily newspapers, business publications, and various electronic media. Credit ratings are also an important part of brokerage reports and commentary on interest-bearing investments. *Barron's* publishes a weekly update on credit-rating changes and America Online subscribers will find listings of S&P and Moody's rating reports in the bond section of the Market News channel. (See Figure 26.)

How Are They Used for Investment Decisions?   Investors can use credit ratings to help match their risk tolerance to their fixed-income investments.

Those investors who prefer high safety to higher yields should choose securities with credit gradings near the top of the scale. Those who can stand some degree of credit risks, often have their risk-taking rewarded with higher yields.

In addition, the overall level of rating changes is seen as one indicator of the nation's economic health. Each agency compiles a periodic report summarizing its recent rating actions. More upgrades than downgrades is a positive signal. When downgrades prevail, it is an ominous sign.

Credit Risk	Moody's	S&P
Prime	Aaa	AAA
Excellent	Aa	AA
Upper medium	A-1, A	A
Lower medium	Baa-1, Baa	BBB
Speculative	Ba	BB
Very speculative	B, Caa	B, CCC, CC
Default	Ca, C	D

Note: Moody's uses the numbers 1 through 3 to denote quality levels within a ranking; S&P uses pluses or minuses.

FIGURE 26   S&P and Moody's Rating Systems

A Word of Caution: Credit risk is only one potential pitfall for fixed-income investors. Too many savers look at credit quality only and ignore another large risk: length of maturity. Price changes owing to changes in interest rates can deeply affect longer-term, fixed-income investments.

Also See: Duration.

Tool #20

Crude Oil Spot Price

What Is This Tool?   The market price for a barrel of unrefined petroleum, typically a high quality grade oil, is called the crude oil spot price. For example, West Texas intermediate crude oil price is the price of a benchmark grade of crude oil produced in West Texas. Crude oil prices are carefully watched as a barometer of everything from global political tensions to inflation to oil company profits. Rising oil prices often boost prices for gasoline, air travel, shipped products, and petroleum-based products. Investors can also invest directly in oil by buying futures and options contracts.

How Is It Computed?   The often-quoted spot price for crude oil is the per barrel price on the most current Light Sweet Crude futures contracts that trade on the New York Mercantile Exchange. Each futures contract calls for delivery of 1000 barrels of oil, but its price is quoted per barrel. The NYMEX, as it is known, also trades futures for sour crude oil, natural gas, and two refined products—unleaded gasoline and heating oil.

Where Is It Found?   The crude oil spot price appears in commodity futures listings in such newspapers as *Barron's, Investor's Business Daily, The New York Times, The Wall Street Journal,* and on computerized databases such as America Online or at quote.yahoo.com. You can also check the NYMEX Web site at www.nymex.com. (See Figure 27.)

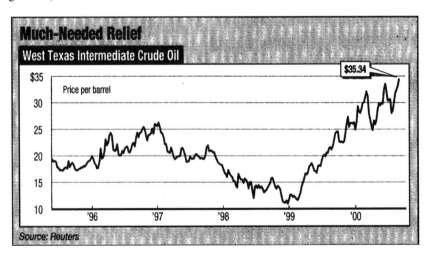

FIGURE 27   How a newspaper covers crude oil prices. (From *Investor's Business Daily,* September 12, 2000, p. A10. With permission.)

How Is It Used for Investment Decisions? Swings in crude oil prices can have a dramatic impact on the world economy and investment markets. Rising oil prices can hurt both stock and bond prices.

Stock investors fear the effect of high oil prices, crippling both to demand because consumers pay more for gasoline and to corporate bottom lines at firms (notably airlines) that are heavy petroleum users. The bond market dislikes any sign of inflation, particularly from oil prices. That is a holdover from the Arab oil embargoes of the 1970s. Inflation erodes the buying power of the cash stream, which is generated by long-term, fixed-income investments.

Rising oil prices can be good for a handful of investments. Oil company stocks enjoy a high-priced environment because profits soar for most petroleum drillers, refiners, and sellers. In addition, short-term cash investments such as money market mutual funds could see their yields improve as higher oil prices likely bump up interest rates.

A Word of Caution: There are major risks associated with investing in commodities futures and options. These securities provide huge leverage—that is, the chance to control a huge amount of petroleum products for a fraction of the cost. But the wrong bet can be financially devastating. Option players can lose only their entire investment while futures speculators can be liable for more than they originally invested if they are far wrong on their hunches.

Tool #21

Currency Indexes

FEDERAL RESERVE TRADE-WEIGHTED DOLLAR

What Is This Tool? The index reflects the currency units of more than 50 percent of the U.S. purchase, principal trading countries.

How Is It Computed? The index measures the currencies of 10 foreign countries: the United Kingdom, Germany, Japan, Italy, Canada, France, Sweden, Switzerland, Belgium, and The Netherlands. The index is weighted by each currency's base exchange rate and, then, averaged on a geometric basis. This weighting process indicates relative significance in overseas markets. The base year was 1973.

Where Is It Found? The index is published by the Federal Reserve System and is found in its *Federal Reserve Bulletin* or at various Federal Reserve Internet sites such as http://woodrow.mpls.frb.fed.us/economy.

How Is It Used for Investment Decisions? The investor should examine the trend in this index to determine foreign exchange risk exposure associated with his or her investment portfolio.

Also, the Federal Reserve trade-weighted dollar is the basis for commodity futures on the New York Cotton Exchange.

Also See: British Pound, German Deutsche Mark.

J.P. MORGAN DOLLAR INDEX

What Is This Tool? The index measures the value of currency units vs. dollars.

How Is It Computed? The index is a weighted-average of 19 currencies including that of France, Italy, the United Kingdom, Germany, Canada, and Japan. The weighting is based on the relative significance of the currencies in world markets. The base of 100 was established for 1980 through 1982.

Where Is It Found? The index appears in *The Wall Street Journal*. (See Figure 28.)

How Is It Used for Investment Decisions? The index highlights the impact of foreign currency units in U.S. dollar terms. The investor can see the effect of foreign currency conversion on U.S. dollar investment.

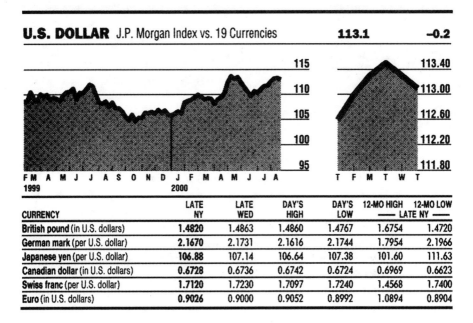

CURRENCY	LATE NY	LATE WED	DAY'S HIGH	DAY'S LOW	12-MO HIGH	12-MO LOW
					── LATE NY ──	
British pound (in U.S. dollars)	1.4820	1.4863	1.4860	1.4767	1.6754	1.4720
German mark (per U.S. dollar)	2.1670	2.1731	2.1616	2.1744	1.7954	2.1966
Japanese yen (per U.S. dollar)	106.88	107.14	106.64	107.38	101.60	111.63
Canadian dollar (in U.S. dollars)	0.6728	0.6736	0.6742	0.6724	0.6969	0.6623
Swiss franc (per U.S. dollar)	1.7120	1.7230	1.7097	1.7240	1.4568	1.7400
Euro (in U.S. dollars)	0.9026	0.9000	0.9052	0.8992	1.0894	0.8904

FIGURE 28 How a newspaper covers foreign currency trading. (From *The Wall Street Journal,* August 25, 2000, p. C1. With permission.)

Tool #22

Dollar-Cost Averaging

What Is This Tool?   Dollar-cost averaging is an investment strategy that attempts to spread out investment risk over time. It does so by requiring small and periodic set dollar amount purchases of an asset over a lengthy period. It is often recommended for small investors who have little investment monies at their disposal or who are unable or unwilling to follow investment markets regularly.

How Is It Computed?   Investors seeking to use this strategy must figure out how much they can afford to contribute systematically to a dollar-cost averaging account each month and/or quarter.

Where Is It Found?   The biggest promoters of dollar-cost averaging have been mutual funds, whose typically small investment minimums allow investors to implement this strategy easily in a cost-effective way. Many funds and brokerages make this process easy by allowing automatic purchases through direct deductions from investors' checking accounts or paychecks.

Investors may unknowingly be using this strategy as part of employer-sponsored savings plans such as 401(k) retirement programs. Many of these benefit plans routinely make equal purchases of assets at set periods, quietly accomplishing dollar-cost averaging.

How Is It Used for Investment Decisions?   Dollar-cost averaging attempts to eliminate one investment decision: timing.

By distributing purchases over a lengthy period, an investor lowers the risk of purchasing a large amount of an investment at the highest possible price. By using a set dollar amount, when prices are high, this strategy tells an investor to buy fewer shares. (See Figure 29.)

When prices are low, the investor would accumulate more discounted shares. Therefore, the strategy screens out whims that could result in the investor buying high and selling low.

Dollar-cost averaging will work as long as prices of the assets targeted by the strategy rise over the long haul.

A Word of Caution: Dollar-cost averaging can result in high transaction costs that can lower returns over time. That is why mutual funds, which often charge either no sales fee or a flat commission, are a popular way to implement this strategy.

Such plans can also create a nightmare at tax time after the investment is sold. Each of the systematic purchases should be separately accounted for on the tax return.

Dollar Cost Averaging

	Dollars Invested	Share Price Paid	Shares Bought
January	$250	$7.00	35.71
February	$250	$6.00	41.67
March	$250	$6.50	38.46
April	$250	$7.25	34.48
May	$250	$6.00	41.67
June	$250	$5.75	43.48
July	$250	$5.25	47.62
August	$250	$5.00	50.00
September	$250	$4.75	52.63
October	$250	$5.75	43.48
November	$250	$6.50	38.46
December	$250	$7.00	35.71
Total dollars invested	$3000	Total shares bought	503.37
			× $7
		Final value	$3523.62

FIGURE 29   A look at how an investor, who spreads out a $3000 investment over equal monthly purchases of $250 each, can profit from dollar-cost averaging when a hypothetical investment sees it price close a year unchanged.

In addition, remember that dollar-cost averaging offers no guarantee of profits. In fact, in bull markets, it will likely cut one's profit because purchases are delayed.

Also See: Value Averaging.

Tool #23

Dow Jones Industrial Average

What Is This Tool?   Dating back to 1885, the Dow Jones Industrial Average or DJIA is the most widely watched benchmark for U.S. stock markets. This index is synonymous with the market's fortunes. When someone says, "The market's up 20," they mean the point change in the Dow index.

How Is It Computed?   The Dow industrial index took its current form of tracking 30 major companies' stocks in October 1928. The 30 stocks as of November 2000 are Allied Signal, Alcoa, American Express, AT&T, Boeing, Caterpillar, Citigroup, Coca-Cola, Disney, du Pont, Eastman Kodak, Exxon Mobil, General Electric, General Motors, Hewlett-Packard, Home Depot, Honeywell, IBM, Intel, International Paper, Johnson & Johnson, McDonald's, Merck, Microsoft, Minnesota Mining & Manufacturing, J.P. Morgan, Philip Morris, Procter & Gamble, SBC, Texaco, United Technologies, and Wal-Mart.

The math behind the index is simple. Each stock's price is divided by the same amount, known as the divisor, to arrive at the index each day. Because the divisor has been less than 1 for some time, share prices are actually multiplied to reach the Dow 30 results. For example, in November 1998, the divisor was 0.25450704. That meant that a $1 movement in a typical DJIA stock would move the index roughly 4 points.

(For trivia buffs, the first 12 members of the Dow in 1885 were American Cotton Oil, American Sugar, American Tobacco, Chicago Gas, Distilling & Cattle Feeding, General Electric, Laclede Gas, National Lead, North American, Tennessee Coal & Iron, U.S. Leather preferred, and U.S. Rubber.)

Where Is It Found?   The Dow 30 is found just about everywhere from virtually every daily newspaper, no matter how small, to reports on television and radio, and can be found on computer services such as America Online. The Dow Jones Web site on the Internet at indexes.dowjones.com has marvelous background on the index. (See Figures 30 and 31.)

How Is It Used for Investment Decisions?   Despite its name recognition, the Dow 30 is seen today as a poor indicator of how the U.S. stock market is doing. Many experts prefer to watch the Standard & Poor's 500 stock index because of its wider reading of big-stock performance. However, because the Dow 30's movement is so widely watched, it cannot be ignored even when it may be giving off incorrect signals about the stock market. The Dow 30 still has a powerful impact on investor psychology. Sharp movements in the Dow 30, regardless of whether they are quickly followed by broader indexes, are closely monitored. For investors who covet the Dow 30, several mutual funds attempt to mimic its performance.

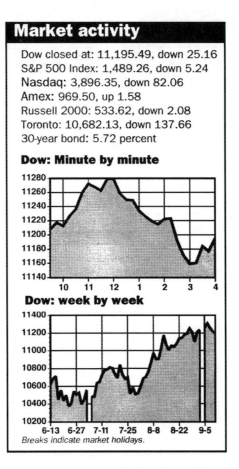

FIGURE 30A   Here's how newspapers cover the Dow Jones Industrial Average. (From *Palm Beach Post,* September 12, 2000, p. 6B. With permission.)

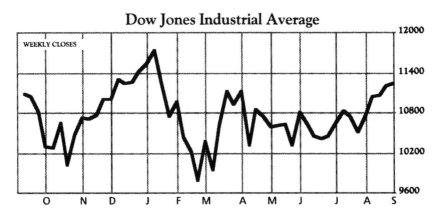

FIGURE 30B   (From *Barron's Market Week,* September 4, 2000, p. MW68. With Permission.)

THE WEEK IN STOCKS

MAJOR INDEXES

12-Month High	Low		Weekly High	Low	Friday Close	Chg.	Weekly %Chg.	12-Month Chg.	%Chg.	Change From 12/31	%Chg.
Dow Jones Averages											
11722.98	9796.03	**30 Indus**	11252.84	11103.01	11238.78	46.15	.41	160.33	1.45	(258.34)	(2.25)
3156.90	2263.59	**20 Transp**	2786.68	2712.92	2712.92	(77.25)	(2.77)	(443.98)	(14.06)	(264.28)	(8.88)
364.98	269.20	**15 Utilities**	364.98	358.41	364.98	9.03	2.54	45.37	14.20	81.62	28.80
3276.26	2751.55	**65 Comp**	3265.01	3223.80	3254.05	8.77	.27	16.97	.52	39.67	1.23
364.71	285.95	**US Tot. Mkt**	358.00	352.85	358.00	5.69	1.62	49.99	16.23	39.67	4.81
509.78	214.58	**Internet**	325.36	300.18	325.36	24.72	8.22	106.10	48.39	(76.40)	(19.02)

FIGURE 30C   (From *Barron's Market Week,* September 4, 2000, p. MW68. With permission.)

Dow Jones Industrial Average Milestones

First day to close above . . .

100	1/12/1906
1,000	11/14/1972
2,000	1/8/1987
3,000	4/17/1991
4,000	2/23/1995
5,000	11/21/1995
6,000	10/14/1996
7,000	2/13/1997
8,000	7/16/1997
9,000	4/6/1998
10,000	3/29/99
11,000	5/3/99

FIGURE 31   Here is a look at key dates in the Dow Jones Industrial Average's history.

   A Word of Caution: The Dow can be actually misleading. Often, the Dow can move in directions that are far different than other big-stock benchmarks, like the S&P 500, as well as benchmarks for small U.S. stocks, like the Russell 2000. Just because the Dow is up, does not mean you have made money. Unless, for example, you own a mutual fund that specifically targets the Dow 30 stocks.

   Also See: Dow Jones Transportation Average, Dow Jones Utilities Average, Standard & Poor's (S&P) Indexes.

Tool #24

Dow Jones Industry Groups

What Are These Tools? Dow Jones publishes statistics of the performance of stocks in industries in various categories: basic materials, conglomerates, consumer/cyclical, consumer/noncyclical, energy, financial, industrial, technology, and utilities. The groupings are further divided into various subcategories. These indexes are further tracked on a global basis: U.S. shares, European, Asia/Pacific, and the Americas.

How Are They Computed? This is a weighted group of industries based on market capitalizations. The base of 100 was established as of June 30, 1982.

Where Are They Found? The performance of industry groups, including those leading and lagging, may be found in *Barron's* and *The Wall Street Journal.*

How Are They Used for Investment Decisions? The investor may examine the index of each major industry, along with 52-week highs and lows, to identify those industries that appear overvalued or undervalued. For example, if the investor concludes that an industry is undervalued and the prospects for that industry are bright, he or she may buy stocks in that industry. Further, industry groups are good to consider as a performance indicator of individual stocks within the industry group. A peer comparison can be quite useful.

A Word of Caution: Industry stock indexes can be misleading. Some companies may be doing well financially even though they are in a depressed industry. (See Figure 32.)

DOW JONES U.S. INDUSTRY GROUPS

Thursday, August 24, 2000 4:00P.M. Eastern Time

Groups Leading (and strongest stocks in group)   **Groups Lagging** (and weakest stocks in group)

GROUPS	CLOSE	CHG	%CHG	GROUP	CLOSE	CHG	%CHG
Biotechnology	**671.35**	**30.88** +	**4.82**	**Pipelines**	**527.13** −	**16.26** −	**2.99**
GeneLogic	22 1/2	4 5/8 +	25.87	ElPasoEngy	57 3/8 −	3 −	4.97
XOMA	7 31/32	1 9/16 +	24.39	Coastal	69 1/8 −	3 3/8 −	4.66
GeronCp	31 11/16	6 3/16 +	24.26	KindrMorgan	36 3/4 −	1 3/8 −	3.61
Consumer Elect	**169.54**	**6.25** +	**3.83**	**Retailers, Apparel**	**230.68** −	**6.17** −	**2.61**
HarmanInt	78	2 7/8 +	3.83	Abercrombie A	22 5/16 −	2 1/8 −	8.70
				Gap Inc	27 1/4 −	1 5/8 −	5.63
				PacSunwear	18 −	7/8 −	4.64
Auto Manufacturers	**463.91**	**15.72** +	**3.51**	**Oil: Majors**	**249.75** −	**5.51** −	**2.16**
GenMotor	72 7/8	3 15/16 +	5.71	USX Marathn	26 11/16 −	1 1/2 −	5.32
FordMotor	27 15/16	1/2 +	1.82	UnocalCp	33 3/4 −	1 3/8 −	3.91
				Texaco	52 5/16 −	1 11/16 −	3.13
Consumer Svc	**730.96**	**23.72** +	**3.35**	**Coal**	**52.19** −	**1.06** −	**1.99**
Travelocity	12 1/2	1 13/16 +	16.96	ConsolEngy	18 5/8 −	3/8 −	2.01
StarMediaNtwk	9 1/16	1 5/16 +	16.94				
About.com	36 37/64	4 3/16 +	12.93				
Footwear	**128.56**	**3.86** +	**3.10**	**Restaurants**	**212.48** −	**3.87** −	**1.79**
TimberInd	43 11/16	2 3/4 +	6.72	CBRL Gp	12 1/8 −	3/8 −	3.00
KenCole A	45 7/8	1 9/16 +	3.53	McDonalds	30 5/16 −	13/16 −	2.61
Reebok	20 1/2	5/8 +	3.14	LoneStarStk	8 5/8 −	3/16 −	2.13

FIGURE 32   How the Dow Jones Industry Groups are covered. (From *The Wall Street Journal,* August 25, 2000, p. C7. With permission.)

Tool #25

Dow Jones Global Stock Indexes

What Is This Tool? This is a grouping of indexes tracking stocks around the globe. The biggest index, the Dow World, tracks shares in 33 countries.

How Is It Computed? The index is based on an equal weighted-average of commodity prices. A 100 base value was assigned to the U.S. index on June 30, 1982; a 100 base was assigned to the rest of the world indexes on December 31, 1991. Indexes are tracked in both local currency and in U.S. dollars, although the dollar tracking contains far more analytical data such as 52-week high and low and year-to-date percentage performance.

Where Is It Found? The index appears daily in *The Wall Street Journal.*

How Is It Used for Investment Decisions? The index can be used to examine the difference between performances in various stock markets and bourses around the globe. That performance can be used to determine if shares in those countries are good values or poor values. In addition, stock indexes often hint at how a country's economy is performing, with rising stock markets typically appearing in healthy economies

A Word of Caution: Currency swings play an important role in calculating investment performance from foreign markets. So if one country's stock market is performing well but its currency is weak vs. the dollar, a U.S. investor may still suffer. Conversely, a country with a strong currency and weak stock market may produce profits for U.S. investors despite the rocky climate for equities. (See Figure 33.)

DOW JONES GLOBAL INDEXES

Region/ Country	DJ Global Indexes, Local Curr. Latest Fri.	Wkly % Chg.	DJ Global Indexes, U.S. $ Latest Fri.	Wkly % Chg.	DJ Global Indexes, U.S. $ on 12/31/99	Point Chg. From 12/31/99	% Chg. From 12/31/99
Americas			354.69	+ 1.19	341.34	+ 13.35	+ 3.91
Brazil	1717293.66	+ 1.52	362.53	+ 1.02	360.26	+ 2.27	+ 0.63
Canada	345.40	+ 0.84	269.03	+ 0.07	207.18	+ 61.85	+ 29.85
Chile	227.76	− 1.37	156.17	− 1.48	173.36	− 17.19	− 9.92
Mexico	457.33	− 3.42	152.00	− 4.28	170.29	− 18.29	− 10.74
U.S.	365.79	+ 1.30	365.79	+ 1.30	354.84	+ 10.95	+ 3.09
Venezuela	459.80	− 3.33	41.26	− 3.23	34.66	+ 6.59	+ 19.02
Latin America			193.65	− 1.45	204.12	− 10.46	− 5.13
Europe/Africa			254.34	− 0.56	275.39	− 21.05	− 7.64
Austria	114.70	− 0.92	80.33	− 1.43	97.98	− 17.65	− 18.02
Belgium	286.92	+ 0.26	201.11	− 0.26	237.32	− 36.21	− 15.26
Denmark	305.45	+ 3.44	218.70	+ 2.86	200.37	+ 18.34	+ 9.15
Finland	1981.45	− 0.42	1245.23	− 0.94	1530.27	− 285.04	− 18.63
France	398.68	+ 0.12	283.98	− 0.40	286.01	− 2.03	− 0.71
Germany	380.38	+ 0.46	265.76	− 0.07	299.78	− 34.02	− 11.35
Greece	545.36	− 3.81	255.93	− 4.34	433.47	− 177.54	− 40.96
Ireland	391.01	− 0.65	263.74	− 1.16	285.83	− 22.08	− 7.73
Italy	431.66	+ 0.89	248.03	+ 0.37	249.55	− 1.52	− 0.61
Netherlands	505.94	− 0.34	353.51	− 0.86	370.40	− 16.89	− 4.56
Norway	241.92	+ 3.24	161.25	+ 2.16	154.78	+ 6.47	+ 4.18
Portugal	377.49	− 0.37	229.52	− 0.88	276.52	− 47.01	− 17.00
South Africa	258.80	+ 1.33	102.66	+ 1.57	115.98	− 13.32	− 11.48
Spain	442.69	− 2.25	233.00	− 2.75	282.86	− 49.87	− 17.63
Sweden	754.28	+ 0.80	449.71	+ 0.19	432.71	+ 16.99	+ 3.93
Switzerland	476.46	+ 0.65	378.25	+ 1.41	369.39	+ 8.86	+ 2.40
United Kingdom	252.76	+ 0.20	198.74	− 1.45	225.98	− 27.24	− 12.06
Pacific Region			111.31	+ 3.15	131.53	− 20.23	− 15.38
Australia	215.17	+ 1.07	162.48	− 1.74	169.96	− 7.48	− 4.40
Hong Kong	363.48	− 3.11	362.44	− 3.12	353.43	+ 9.00	+ 2.55
Indonesia	173.17	− 1.97	41.41	− 1.97	71.72	− 30.31	− 42.26
Japan	87.43	+ 3.33	102.15	+ 5.39	124.10	− 21.95	− 17.69
Malaysia	149.81	− 2.52	107.27	− 2.52	111.77	− 4.49	− 4.02
New Zealand	132.78	− 4.10	107.02	− 7.17	141.39	− 34.37	− 24.31
Philippines	135.65	+ 0.26	78.38	+ 0.15	130.59	− 52.20	− 39.98
Singapore	152.63	− 1.13	143.83	− 1.16	170.56	− 26.73	− 15.67
South Korea	130.73	+ 0.11	88.97	+ 0.20	128.97	− 40.00	− 31.01
Taiwan	197.17	− 1.83	164.08	− 1.71	175.45	− 11.36	− 6.48
Thailand	58.55	− 3.10	33.84	− 2.53	55.58	− 21.74	− 39.11
Europe/Africa (ex. South Africa)			263.12	− 0.60	284.68	− 21.56	− 7.57
Europe/Africa (ex. U.K. & S. Africa)			302.97	− 0.22	321.58	− 18.61	− 5.79
Nordic Region			430.90	+ 0.20	455.76	− 24.87	− 5.46
Pacific Region (ex. Japan)			173.13	− 1.99	191.87	− 18.75	− 9.77
World (ex. U.S.)			176.14	+ 0.70	193.65	− 17.51	− 9.04
DOW JONES WORLD STOCK INDEX			245.52	+ 1.00	253.77	− 8.26	− 3.25

Indexes based on 12/31/91 = 100.　　　　　　　　　

FIGURE 33　How the Dow Jones Global Indexes are covered. (From *Barron's Market Week,* August 28, 2000, p. MW8. With permission.)

Tool #26

Dow Jones Transportation Average

What Is This Tool?   This is a widely watched benchmark for the movement of U.S. transportation stocks. These issues are viewed by many analysts as highly cyclical; thus, their performance can be seen as an indicator of future economic activity.

How Is It Computed?   The Dow transportation index comprises 20 major transportation companies' stocks—from airlines to railroads. As of November 2000, the list was comprised of AMR, Airborne Freight, Alexander Baldwin, Burlington Northern, CNF, CSX, Delta Air, FDX (Federal Express), GATX, JB Hunt, Norfolk Southern, Northwest Air, Roadway, Ryder Systems, Southwest Air, UAL, US Airways, U.S. Freightways Union Pacific, and Yellow Corp. Each stock's price is multiplied by the same fixed amount to arrive at the index each day. That means that movement in any of those 20 stocks will have an equal impact on the index's final tally.

Where Is It Found?   The Dow Transportation Average is published in most daily newspapers such as *The Los Angeles Times* and *Washington Post* with listings of its better known sister index, the Dow Jones Industrial Average. *Barron's* includes detailed information on the index and its component stocks. It also can be found on computer services such as America Online or at the Dow Internet site at indexes.dowjones.com. (See Figure 34.)

How Is It Used for Investment Decisions?   The Dow Transportation Average can be used to weigh how the stock market values transportation-related issues. A rising index can signal investors' sentiment swinging toward these cyclical transportation issues. Conversely, a falling index can be viewed as a negative sign. The index can be used loosely to track oil prices, because fuel is a major component of airline profitability. When the Dow Transports are up, it is a good bet that there is an outlook for lower oil prices.

The Dow Transportation Average is also a component of the broader Dow Jones Composite Index, which also includes shares in the Dow Industrial and Dow Utilities Index. The composite index is seen by some analysts as a good indicator for the general market's success.

A Word of Caution: These stocks were once seen as a key indicator of the health of the U.S. economy. However, the age of deregulation has beaten down many of these companies and their share prices. Some analysts now question whether this index can be used as a barometer of future economic activity.

Also See: Dow Jones Industrial Average, Dow Jones Utilities Average.

DJ TRANSPORTATION AVERAGE

The weekly Dow Transports; with a high/low range based upon the daily closing average.

Week Ended		First	High	Low	Last		Chg.
2000							
Aug	25	2818.59	2823.08	2735.35	2790.17	–	46.97
	18	2921.21	2921.21	2837.14	2837.14	–	90.36
	11	2894.21	2927.50	2847.91	2927.50	+	40.69
	4	2853.09	2887.11	2858.09	2886.81	+	117.28
Jul	28	2820.09	2820.09	2769.53	2769.53	–	38.38
	21	2900.27	2900.27	2794.06	2808.42	–	110.62
	14	2809.02	2919.04	2809.02	2919.04	+	134.40
	7	2704.83	2806.63	2704.83	2784.64	+	139.27
Jun	30	2614.55	2719.86	2614.55	2645.37	+	14.66
	23	2680.67	2680.67	2612.76	2630.71	–	42.48
	16	2781.50	2781.50	2673.19	2673.19	–	116.98
	9	2800.34	2800.34	2750.68	2790.17	–	39.19
	2	2741.70	2829.36	2709.99	2829.36	+	141.81
May	26	2720.16	2829.07	2687.55	2687.55	–	54.45
	19	2838.04	2865.27	2742.00	2742.00	–	132.02
	12	2935.87	2935.87	2863.17	2874.02	–	2.09
	5	2858.68	2876.11	2794.96	2876.11	+	26.10
Apr	28	2805.13	2914.03	2805.13	2850.01	+	16.76
	21	2678.58	2833.25	2678.58	2833.25	+	106.21
	14	2843.13	2952.63	2727.04	2727.04	–	100.68
	7	2706.10	2847.32	2706.10	2827.72	+	64.48

FIGURE 34   How the Dow Jones Transportation Average is covered. (From *Barron's Market Week*, August 18, 2000, p. MW74. With permission.)

Tool #27

Dow Jones Utilities Average

What Is This Tool?   This is a closely viewed benchmark for U.S. utility stocks, which are a popular investment for conservative, income-oriented investors. Both its price movement and the yield on the average are watched by investors.

How Is It Computed?   The Dow Utilities Index comprises shares in 15 large publicly owned utilities that provide consumers with everything from gas to water to electricity. As of November 2000, the index was comprised of American Electric, Columbia Gas, Con Edison, Dominion Resources, Duke Energy, Edison International, Enron, PG & E Corp., PECO, Public Service Enterprises, Southern Co., TXU, Unicom, and Williams Cos.

Each stock's price is multiplied by the same amount to arrive at the index each day. That means that movement in any of those 15 stocks will have an equal impact on the index's result.

The Dow Utility Index's yield is found by totaling all dividends paid by the companies comprising the index and dividing that sum by the combined share price of the 15 utilities.

Where Is It Found?   Daily price changes in the Dow Utilities Index can be located in most daily newspapers such as *The New York Times* and *The Wall Street Journal.* It is usually found alongside listings of its better known sister index, the Dow Jones Industrial Average. The index's yield is somewhat more difficult to locate. It also can be found on computer services such as America Online or at the Dow Internet site at indexes.dowjones.com. (See Figure 35.)

How Is It Used for Investment Decisions?   The Dow Utilities Index can be used to evaluate the direction of utility stocks and how their yield compares to other income-oriented investments.

A rising index is a signal that investors are buying utility stocks. This occurs notably in two situations. First, when interest rates are falling and utility stock yields become attractive. Second, when the economy is picking up, translating to higher sales to industrial users of utility services.

Conversely, a falling index can be a sign that investors fear that interest rates will rise or that a weak economy may slow industrial outputs.

The Utilities Index's yield can be used to gauge the overall attractiveness of such stocks to an income-oriented investor. In addition, it can help measure how a single utility's dividend payout compares to its peers.

A Word of Caution: Utility prices are not moved by national economics only. Other items that impact utility share prices are regulatory and environmental

DOW JONES UTILITIES

The table below lists the total earnings (losses) of the Dow Jones Utility Average component stocks of record based upon generally accepted accounting principles as reported by the company and adjusted by the Dow Divisor in effect at quarter end and the total dividends of the component stocks based upon the record date and adjusted by the Dow Divisor in effect at quarter end. N.A.-Not available.

Year Ended	Quarter Ended		Close Avg.	12-Mth. Earns	P/E Ratio	Qtrly Divs.
2000	June	30	306.91	23.21	13.2	3.18
	Mar.	31	291.77	22.13	13.2	3.18
1999	Dec.	31	283.36	18.60	15.2	2.95
	Sept.	30	298.26	18.46	16.2	2.94
	June	30	316.82	15.14	20.9	2.80
	Mar.	31	292.28	15.44	18.9	2.79
	Yr.End					11.48
1998	Dec.	31	312.30	15.42	20.2	2.78
	Sept.	30	306.72	14.20	21.6	2.76
	June	30	293.87	13.96	21.0	2.76
	Mar.	31	285.94	12.97	22.0	2.66
	Yr.End					10.96
1997	Dec.	31	273.07	13.79	19.8	2.75
	Sept.	30	238.37	14.76	16.1	2.66

FIGURE 35 How a newspaper covers the Dow Jones Utility Average. (From *Barron's Market Week,* August 18, 2000, p. 83. With permission.)

concerns, factors that have less significance to the broad investment outlook. An investor must determine if such noneconomic factors are figuring into the index's movement before making an investment based on a Dow Utilities Average's trend.

Also See: Dow Jones Industrial Average, Dow Jones Transportation Average, Economic Indicators, and Bond Yields.

Tool #28

Duration

What Is This Tool? Duration is a way to measure the risk of price change owing to interest rate fluctuations in a bond or a portfolio of bonds. While rarely discussed in the media, it is the top figure that professional money managers watch when reviewing a bond portfolio.

How Is It Computed? Duration is a complex calculation that includes evaluating the income stream that a bond or a bond portfolio generates. That cash flow is then discounted, creating a present value of that interest payment stream. The calculation also includes an estimate of the chances for the bond or bonds to be called back by the issuer.

As an example, a 30-year U.S. Treasury bond has a duration of approximately 10 years.

The results are stated as figures in years. Simply translated, that means that for each year of duration, bonds or portfolio will lose or gain 1 percent of principal value for each 1 percentage point move in interest rates.

As an example, an investor who owns a 30-year U.S. Treasury (duration 10 years) would lose 15 percent of his principal if 30-year interest rates were to rise 1.5 percentage points. This calculation is somewhat subjective and, sometimes, can misstate the bondholders' risk if certain incorrect assumptions are used. (See Figure 36.)

Where Is It Found? Duration is the most common measure of interest-rate risk used by bond traders. Typically, a bond's stated maturity or a portfolio's average maturity (a simple calculation of weighing bond maturity and prices) is quoted. Such maturity figures do not as clearly outline the interest risks involved.

Owners of bond portfolios, such as mutual funds, probably can get duration figures from the management company. Various reports by the fund tracker *Morningstar Inc.* include duration figures for bond mutual funds.

How Is It Used for Investment Decisions? Duration is a powerful tool that shows investors how much price risk exists in holding longer-term bonds just from interest rate swings. Bond prices fall as rates rise and values increase when rates drop. The price swings can be dramatic.

An investor comparing two bond portfolios with equal yields but different durations might choose the one with the longer duration if he believes that interest rates were going to fall. That portfolio would likely produce more capital gains if rates did go lower. However, if an investor thought rates were going to rise, or the investor

McCaulay's Duration

Year	Income Stream	Present Value (PV) of Income Stream	PV as a Percent of Bond's Face Value	Year × Percent
1	$ 65	$61.03	6.1%	0.061
2	$ 65	$57.31	5.7%	0.115
3	$ 65	$53.81	5.4%	0.161
4	$ 65	$50.53	5.1%	0.202
5	$ 65	$47.44	4.7%	0.237
6	$ 65	$44.55	4.5%	0.267
7	$ 65	$41.83	4.2%	0.293
8	$ 65	$39.28	3.9%	0.314
9	$ 65	$36.88	3.7%	0.332
10	$ 1,065	$567.35	56.7%	5.674
				Sum of years equals duration 7.656

FIGURE 36 McCaulay's duration is a popular method. First, the present value of a bond's income stream must be determined. Second, that present value must be calculated as a percentage of the bond's price. Finally, that percentage must be multiplied by the corresponding year's number. The sum of the products equals the duration.

simply wanted to lower risk-taking, he or she should choose the portfolio with the shorter duration. (See Figure 37.)

A Word of Caution: Duration is not a static figure. Movements in interest rates alone will change a portfolio's duration. It will be further changed if a manager then takes actions in response to market movements. This means that an investor relying on duration to watch, for example, a bond mutual fund must make sure that the information is up-to-date. If rates were rising, an investor might want to prune a long-duration fund from his portfolio. However, the investor would first want to check to see if the fund's manager had already taken defensive moves and lowered the fund's duration.

Also See: Economic Indicators and Bond Yields, Yield on a Bond, Yield on an Investment: Current.

Mutual Fund Duration

Fund Name	Duration	5-Year Total Return (Annualized)	10-Year Total Return (Annualized)	Total Assets (Billions)
Pimco Total Return	5.0	8.0%	10.4%	$ 18.50
Franklin California Tax-Free Income	6.0	6.3%	7.9%	$ 14.98
Vanguard F/I GNMA	3.6	7.4%	9.1%	$ 10.16
Bond Fund of America	5.1	6.4%	9.1%	$ 9.18
Franklin U.S. Government Securities I	2.9	6.8%	8.5%	$ 8.92
Vanguard Muni Intermediate-Term	5.2	5.7%	7.9%	$ 7.50
Franklin Fed Tax-Free Income I	5.3	6.2%	8.1%	$ 7.08
Vanguard Bond Index Total	4.4	7.1%	9.1%	$ 6.62
Franklin High Yield T/F Inc I	6.8	7.4%	8.7%	$ 6.04
IDS High-Yield Tax-Exempt A	6.5	5.9%	7.7%	$ 5.74
Vanguard Short-Term Corp	2.2	6.2%	7.9%	$ 5.20
MSDW U.S. Government Securities B	4.9	6.2%	7.5%	$ 5.17
Vanguard High-Yield Corporate	4.8	8.9%	9.7%	$ 4.95
Franklin New York Tax-Free Income I	5.5	6.1%	8.2%	$ 4.83
Fidelity Spartan Muni Income	7.1	5.7%	8.1%	$ 4.63
MAS Fixed-Income	4.5	7.1%	9.7%	$ 4.54
AARP GNMA and U.S. Treasury	2.8	6.0%	7.8%	$ 4.54
Oppenheimer Strategic Income B	4.8	6.3%	—	$ 4.04
Oppenheimer Strategic Income A	4.8	7.1%	—	$ 3.95
Vanguard Long-Term Corporate	9.4	8.4%	11.0%	$ 3.95
Standish Fixed-Income	4.8	6.8%	9.5%	$ 3.46
Fidelity Intermediate Bond	3.4	5.8%	8.1%	$ 3.45
Dreyfus Municipal Bond	7.1	5.1%	7.6%	$ 3.35
Rochester Fund Municipals A	6.5	6.2%	8.8%	$ 3.34
Kemper U.S. Government Securities A	3.2	6.6%	8.4%	$ 3.34

FIGURE 37 Here is a look at duration figures for 25 major bond mutual funds. All data is as of September 30, 1998. (From Morningstar Inc. With permission.)

Tool #29

Economic Indicators and Bond Yields

What Are These Tools? The bond investor makes an analysis of the economy primarily to determine his or her investment strategy. It is not necessary for the investor to formulate his or her own economic forecasts. The investor can rely on published forecasts in an effort to identify the trends in the economy and adjust his or her investment position accordingly.

The investor must keep abreast of the economic trend and direction and attempt to see how they affect bond yields and bond prices. Unfortunately, there are too many economic indicators and variables to be analyzed. Each has its own significance. In many cases, these variables could give mixed signals about the future of the economy and, therefore, mislead the investor.

How Are They Computed? Various government agencies and private firms tabulate the appropriate economic data and calculate various indices.

Where Are They Found? Sources for these indicators are easily subscribed at an affordable price or can be found in your local public and college libraries. They include daily local newspapers and national newspapers such as *USA Today, The Wall Street Journal, Investor's Business Daily, Chicago Tribune* and *The New York Times* or periodicals, such as *Business Week Forbes, Fortune, Money, Kiplinger's Personal Finance Magazine, Worth, Barron's, Smart Money, Nation's Business,* and *U.S. News and World Report.* Internet users could check out the semiannual Livingston Survey, started in 1946 by the late economist Joseph A. Livingston. It is the oldest continuous survey of economists' expectations. The Federal Reserve Bank of Philadelphia took responsibility for the survey in 1990 and it can be found at the bank's Web site at http://www.phil.frb.org/econ/liv/.

How Are They Used for Investment Decisions? Figure 38 provides a concise and brief list of the significant economic indicators and how they affect bond yields. Remember, bond yields and bond prices act conversely, so a rise in yields means a fall in prices and vice versa.

A Word of Caution: Figure 38 merely serves as a handy guide and should not be construed as an accurate predictor in all cases. Many times, the anticipation of good or bad news is built into the market and when the news comes out, the reverse move happens. That is because traders are unwinding the positions they took to profit from that news.

Probable Effects of Economic Variables on Bond Yields*

Indicators[†]	Effects on Bond Yields[‡]	Reasons
GNP and industrial production falls	Fall	As economy slows, Fed may ease credit by allowing rates to fall
Unemployment rises	Fall	High unemployment indicates lack of economic expansion; Fed may loosen credit
Inventories rise	Fall	Inventory levels are good indicators of economic slowdown
Trade deficit rises	Rise	Dollar weakens, that is inflationary
Leading indicators rise	Rise	Advance signals about economic rise health; Fed may tighten credit
Housing starts rise	Rise	Growing economy owing to increased new housing demand; Fed may tighten; mortgage rates rise
Personal income rises	Rise	Higher income means higher consumer spending, thus inflationary; Fed may tighten
Inflation		
Consumer Price Index up	Rise	Inflationary fears rise
Producer Price Index up	Rise	Early signal for inflation increase
Monetary Policy		
Money supply rises	Rise	Excess growth in money supply is inflationary; Fed may tighten
Fed fund rate rises	Rise	Increase in business and consumer loan rates; used by Fed to slow economic growth and inflation
Fed buys (sells) bills	Rise (fall)	Adds (deducts) money to the economy; interest rates may go down (up)
Required reserve rises	Rise	Depresses banks' lending

*This table merely serves as a handy guide and should not be construed as accurate at all times.

†Fall in any of these indicators will have the opposite effect on bond yields.

‡Note that the effects are based on yield and are, therefore, opposite of how bond prices will be affected.

FIGURE 38   Here is how economic news and bond market reaction, at times, mix.

Also See: Economic Indicators and the Stock Market.

Tool #30

Economic Indicators and Stocks

What Is This Tool? The investor makes an analysis of the economy primarily to determine his or her investment strategy. It is not necessary for him or her to formulate his or her own economic forecasts. The investor can rely on published forecasts in an effort to identify the trends in the economy and adjust his or her investment position accordingly. The investor must keep abreast of the economic trend and direction and attempt to see how it affects the security market. Unfortunately, there are too many economic indicators and variables to be analyzed. Each has its own significance. In many cases, these variables could give mixed signals about the future of the economy and, therefore, mislead the investor.

How Is It Computed? Various government agencies and private firms tabulate the appropriate economic data and calculate various indices.

Where Is It Found? Sources for these indicators are easily subscribed at an affordable price or can be found in your local public and college libraries. They include daily local newspapers and national newspapers such as *USA Today, The Wall Street Journal, Investor's Business Daily, The Los Angeles Times* and *The New York Times* and periodicals such as *Business Week* (See Figure 39), *Forbes, Fortune, Money, Kiplinger's Personal Finance Magazine, Worth, Barron's, Smart Money, Nation's Business,* and *U.S. News and World Report.* Internet users can look at the White House Web site's Economic Statistics Briefing Room that provides easy access to current Federal economic indicators. The Briefing Room is at www.whitehouse. gov/fsbr/esbr.html.

How Is It Used for Investment Decisions? Figure 39 summarizes the types of economic variables and their probable effect on the security market and the economy, in general.

A Word of Caution: The accompanying chart merely serves as a handy guide and should not be construed as an accurate predictor in all cases. Many times the anticipation of good or bad news is built into the market and when the news comes out, the reverse move happens. That is because traders are unwinding the positions they took to profit from that news.

Also See: Economic Indicators and Bond Yields.

Economic Variables and Their Impacts on the Economy and Stocks

Economic Variables	Impact on Stock Market
Real growth in GNP	Positive (without inflation) for stocks
Industrial production down	Consecutive drops are a sign of recession; bad for stocks
Inflation	Detrimental to stocks
Flat pricing	Great for stocks
Deflation	Bad for stocks
High capacity utilization	Can be positive, but full capacity is inflationary
Falling durable goods orders	Consecutive drops are a sign of recession; very bad for stocks in cyclical industries
Increase in business investment	Positive for stocks, especially capital goods makers
Increase in consumer confidence, personal income	Positive for stocks, especially retailers
Falling leading indicators	Bearish for stocks; drops are a sign of bad times ahead
Rising housing starts	Positive for housing stocks
Rising corporate profits	Positive for stocks; corporate bonds also fare well
Unemployment up	Upward trend unfavorable for stocks and economy
Increase in business inventories	Positive for those fearful of inflation; negative for those looking for growing economy
Lower federal deficit	Lowers interest rates, good for many stocks; potential negative for depressed economy
Deficit in trade and balance of payments	Negative for economy and stocks of companies facing stiff import competition
Weak dollar	Inflationary for economy; good for companies with stiff foreign competition
Rising interest rates	Can choke off investment in new plants and lure skittish investors from stocks

FIGURE 39   Here is how economic news and stock market reaction mix, at times.

Tool #31

Economic Indicators: Factory Orders and Purchasing Manager's Index

What Are These Tools? The factory order series presents new orders received by manufacturers of durable goods other than military equipment. (Durable goods are defined as those having a useful life of more than three years.) Nondefense equipment represents about $\frac{1}{5}$ to $\frac{1}{3}$ of all durable goods production. The series includes engines, construction, mining, and materials handling equipment; office and store machinery; electrical transmission and distribution equipment and other electrical machinery (excluding household appliances and electronic equipment); and railroad, ship and aircraft transportation equipment. Military equipment is excluded because new orders for such items do not respond directly to the business cycle.

The National Association of Purchasing Management releases its monthly *Purchasing Index* which tells you about buying intentions of corporate purchasing agents.

How Is It Computed? The factory order series is released by the Department of Commerce. Each month, more than 2000 companies are asked to file a report covering orders, inventories, and shipments.

As for the Purchasing Index, the National Association of Purchasing Agents conducts a survey that polls purchasing managers from key industries. (See Figure 40.)

Where Is It Found? The factory order series is reported in daily newspapers and business dailies. Commerce Department statistics can be found at www.census.gov/econ/ on the Internet.

How Is It Used for Investment Decisions? Economists typically count on factory production, particularly of "big ticket" durable goods ranging from airplanes to home appliances, to help lift the economy from a downturn. A decline in this series suggests that factories are unlikely to hire new workers. A drop in the backlog of unfilled orders is also an indication of possible production cutbacks and layoffs. The wider dispersal of gains in many types of goods is looked upon as a favorable sign for economic recovery. The broader the dispersal of order increases, the broader is rehiring.

The purchasing managers are responsible for buying the raw materials that feed the nation's factories. Their buying patterns are considered a good indication of the

Factory orders in steep decline

Orders to U.S. factories posted their biggest-ever one-month drop, further proof that the economy is cooling, economists said.

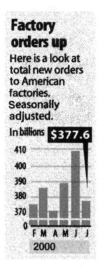

Factory orders up

Here is a look at total new orders to American factories. Seasonally adjusted.

In billions **$377.6**

410
400
390
380
370
0

F M A M J J
2000

The Commerce Department said Thursday that factory orders plunged a record 7.5 percent in July as demand fell sharply for airplanes, electronics and communications equipment.

The Commerce Department reported Thursday that total factory orders fell to a seasonally adjusted $377.6 billion in July, down from $408.1 billion the month before.

The drop was bigger than the 6.8 percent decline many analysts expected.

FIGURE 40   Here's how factory orders are tracked by a newspaper. (From *Sun Sentinel,* September 1, 2000, p. D1. With permission.)

direction of the economy. A reading of 50 percent or more percent indicates that the manufacturing economy is generally expanding. A reading above 44.5 percent over a period of time indicates that the overall economy is augmenting.

A Word of Caution: Again, in order to make an overall assessment of the economy, the investor must look to other important economic indicators.

Also See: Economic Indicators and the Security Market, Economic Indicators and Bond Yields.

Tool #32

Economic Indicators: Gross Domestic Product

What Is This Tool? Gross Domestic Product (GDP) measures the value of all goods and services produced by the economy within its boundaries and is the nation's broadest gauge of economic health. GDP is normally stated in annual terms, although data are compiled and released quarterly. (See Figure 41.)

How Is It Computed? The Department of Commerce compiles the GDP. It is reported as a "real" figure, that is, economic growth minus the impact of inflation. The figure is tabulated on a quarterly basis, coming out in the month after a quarter has ended. It is then revised at least twice, with those revisions being reported once in each of the months following the original release.

Economic Calendar								Oct 02 - Oct 06

Last Week   Next Week

Date	Time (ET)	Statistic	For	Actual	Briefing Forecast	Market Expects	Prior	Revised From
Oct 02	12:00 am	Auto Sales	Sep	-	6.7M	6.7M	6.8M	-
	12:00 am	Truck Sales	Sep	-	7.5M	7.6M	7.8M	-
	10:00 am	Construction Spending	Aug	1.4%	0.2%	0.5%	-1.9%	-1.6%
	10:00 am	NAPM Index	Sep	49.9%	50.7%	50.0%	49.5%	49.5%
Oct 03	9:00 am	FOMC Meeting	---	-	---	---	-	-
	10:00 am	Leading Indicators	Aug	-0.1%	-0.1%	-0.1%	-0.2%	-0.1%
	10:00 am	New Home Sales	Aug	893K	875K	895K	921K	944K
Oct 04	10:00 am	Factory Orders	Aug	-	1.8%	1.7%	-7.9%	-
	10:00 am	NAPM Services	Sep	-	60.0%	60.5%	60.0%	-
Oct 05	8:30 am	Initial Claims	09/30	-	305K	300K	287K	-
	2:00 pm	FOMC Minutes	08/22	-	-	-	-	-
Oct 06	8:30 am	Average Workweek	Sep	-	34.4	34.4	34.3	-
	8:30 am	Hourly Earnings	Sep	-	0.3%	0.3%	0.3%	-
	8:30 am	Nonfarm Payrolls	Sep	-	215K	225K	-105K	-
	8:30 am	Unemployment Rate	Sep	-	4.1%	4.1%	4.1%	-
	3:00 pm	Consumer Credit	Aug	-	$10.0B	$10.0B	$9.4B	-

Key: Better or Worse than market expects.

FIGURE 41A Gross Domestic Product. Here's how investors can track upcoming economic indicators. (Reproduced with permission of Yahoo! Inc. © 2000 by Yahoo! Inc. YAHOO! and the YAHOO! Logo are trademarks of Yahoo! Inc.)

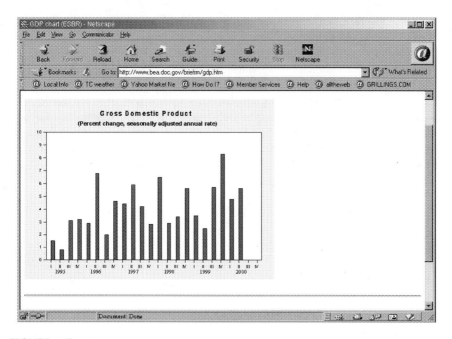

FIGURE 41B Gross Domestic Product. Here's how investors can watch the gross domestic product as compiled by the U.S. government at http://www.whitehouse.gov/fsbr/esbr.html.

Where Is It Found? GDP reports appear in most daily newspapers and online at services like America Online. An investor may also visit the Federal Government Statistics Web site on the Internet at http://www.fedstats.gov/.

How Is It Used for Investment Decisions? GDP is often a measure of the state of the economy. For example, many economists speak of recession when there has been a decline in GDP for two consecutive quarters. The GDP in dollar and real terms is a useful economic indicator. An expected growth rate of 3 percent in real terms would be very attractive for long-term investment and would affect the stock market positively. Because inflation and price increases are detrimental to equity prices, a real growth of GDP without inflation is favorable and desirable.

The following diagram charts a series of events leading from a rising GDP to higher security prices.

GDP up → Corporate profits up → Dividends up → Stock prices up

Generally speaking, too much growth is inflationary and, thus, negative for the stock and bond markets. When companies are producing "flat out," they need workers desperately and are willing to pay big wage increases to attract new workers and keep them. These wage increases, however, raise business costs and lead firms to raise prices and must be avoided. Too little production is undesirable as well. Low levels of production mean layoffs, unemployment, low incomes for workers, and tend to depress the stock market.

Investors watching for signs of inflation should check the "deflator" portion of the GDP report. That contains what some experts feel is the most detailed tracking of price pressures from the government.

A Word of Caution: GDP fails the timely release criterion for useful economic indicators. Unfortunately, there is no way of measuring whether we are in a recession or prosperity currently, based on the GDP measure. Only after the quarter is over, can it be determined if there was growth or decline. Experts look upon other measures such as unemployment rate, industrial production, durable orders, corporate profits, retail sales, and housing activity to look for a sign of recession.

Also See: Economic Indicators and Bond Yields, Economic Indicators and the Security Market, Economic Indicator: Recession.

Tool #33

Economic Indicators: Housing Starts and Construction Spending

What Are These Tools?   Housing starts is an important economic indicator followed by investors and economists that offer an estimate of the number of dwelling units on which construction has begun during a stated period. It covers construction of new homes and apartments. When an economy is going to take a downturn, the housing sector is the first to decline. This indicates the future strength of the housing sector of the economy. At the same time, it is closely related to interest rates and other basic economic factors.

The statistics for construction spending covers homes, office buildings, and other construction projects.

How Are They Computed?   Both housing starts and construction spending figures are issued monthly by the Department of Commerce. Visit the Federal Government Statistics Web site at http://www.fedstats.gov/.

Where Are They Found?   National business daily newspapers and many local newspapers report on these property-related figures as does any reliable Internet-based financial news service. (See Figure 42.)

How Are They Used for Investment Decisions?   Housing is a key interest-sensitive sector that usually leads the rest of the economy out of the recession. Also, housing is vital to a broader economic revival, not only because of its benefits for other industries but also because it signals consumers' confidence about making long-term financial commitments.

A Word of Caution: For the housing sector to be sustained, housing start figures need to be backed by building permits. Permits are considered a leading indicator of housing starts.

Also See: Economic Indicators and Bond Yields, Economic Indicators: Interest Rates, Economic Indicators and the Security Market.

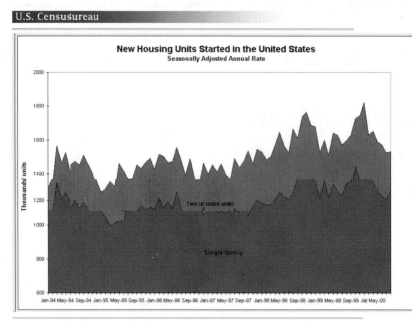

Source: U.S. Census Bureau

FIGURE 42   How a government web site charts housing-start figures.

Tool #34

Economic Indicator: Index of Leading Indicators

What Is This Tool?   The Index of Leading Indicators is the economic series of indicators that tends to predict future changes in economic activity. This index was designed to reveal the direction of the economy in the next six to nine months. By melding 10 economic yardsticks, an index is created that has shown a tendency to change before the economy makes a major turn. Hence, the term "leading indicators." The index is designed to forecast economic activity six to nine months ahead.

How Is It Computed?   This series is calculated and published monthly by the Conference Board and consists of the following.

- Average weekly hours for U.S. manufacturing workers—employers find it a lot easier to increase the number of hours worked in a week than to hire more employees.
- Average weekly initial claims for unemployment insurance—the number of people who sign up for unemployment benefits signals changes in present and future economic activity.
- Manufacturers' new orders, consumer goods and materials—new orders mean more workers hired, more materials and supplies purchased, and increased output. Gains in this series usually lead recoveries by as much as four months.
- Vendor performance, slower deliveries diffusion index—represents the percentage of companies reporting slower deliveries. As the economy grows, firms have more trouble filling orders.
- Manufacturers' new orders, nondefense capital goods—factories will employ more as demand for big-ticket items, especially those not bought by the government, stay strong.
- Building permits, new private housing units—optimistic builders is often a good sign for the economy.
- Stock prices, 500 common stocks—stock market advances usually precede business upturns by three to eight months.
- Money supply, M2—A rising money supply means easy money that sparks brisk economic activity. This usually leads recoveries by as much as 14 months.

- Interest rate spread, 10-year Treasury bonds minus federal funds rate—steep yield curve, when long rates are much higher than short ones, is sign of healthy economic outlook.
- Consumer expectations index—Consumer spending buys two-thirds of GDP (all goods and services produced in the economy), so any sharp change could be an important factor in an overall turnaround.

Where Is It Found?   The monthly report is well covered by daily business publications, major newspapers, business TV shows, and on the Internet. You can also check the Conference Board's Web site at www.conference-board.org.

How Is It Used for Investment Decisions?   If the index is consistently rising, even only slightly, the economy is chugging along and a setback is unlikely. If the indicator drops for three or more consecutive months, look for an economic slowdown and possibly a recession in the next year or so.

A rising (consecutive percentage increases in) indicator is bullish for the economy and the stock market, and vice versa. Falling index results could be good news for bondholders looking to make capital gains from falling interest rates.

Now the Conference Board points out that while it is often stated in the press that three consecutive downward movements in the leading index signal a recession, they do not endorse the use of such a simple, inflexible rule. Their studies show that a 1 percent decline (2 percent when annualized) in the leading index, coupled with declines in a majority of the 10 components, provides a reliable, but not perfect, recession signal.

A Word of Caution: The composite figure is designed to tell only in which direction business will go. It is not intended to forecast the magnitude of future ups and downs. The index has also given some false warning signals in recent years.

Also See: Economic Indicators and Bond Yields, Economic Indicators and the Stock Market.

TOOL #35

ECONOMIC INDICATORS: INDUSTRIAL PRODUCTION AND CAPACITY UTILIZATION

What Is This Tool? The index of industrial production, more precisely, the Federal Reserve Board Index of Industrial Production, measures changes in the output of the mining, manufacturing, and gas and electric utilities sectors of the economy. Detailed breakdowns of the index provide a reading on how individual industries are faring.

Industrial production is narrower than gross domestic product (GDP) because it omits agriculture, construction, wholesale and retail trade, transportation, communications, services, finance, and government.

Another way to view the performance of the real economy is to look at industrial production relative to the production capacity of the industrial sector. The actual production level as a percent of the full capacity level is called the rate of capacity utilization. This monthly rate is limited to manufacturing industries.

How Is It Computed? Data for the index is drawn from 250 data series obtained from private trade associations and internal estimates.

Where Is It Found? This monthly Index of Industrial Production is released two weeks into the next month and is published by the Federal Reserve Board. The rate of capacity utilization is announced every month by the Fed, one day after the Index of Industrial Production. Both are published in the *Federal Reserve Bulletin* and appear in major daily newspapers and on online computer news services such as America Online. (See Figure 43.)

How Is It Used for Investment Decisions? As the index rises, this is a sign that the economy will strengthen and that the stock market should turn up. A falling industrial production should be a concern for the economy and the investor. Regardless of the state of the economy, however, detailed breakdowns of the index provide a reading on how individual industries are faring and on what industries should be attended by investors.

A rising rate of capacity utilization is positive for the economy and the stock market; a falling rate is an indication of a sinking economy and, thus, negative for the stock market.

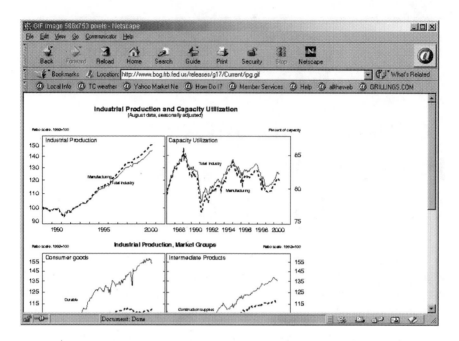

FIGURE 43   Here's a look at how a government web site charts industrial production.

A Word of Caution: Industrial production is more volatile that GDP, because GDP, unlike industrial production, includes activities that are largely spared cyclical fluctuations, such as services, finance, and government.

Also See: Economic Indicators, Factory Orders and Purchasing Manager's Index, Economic Indicators and Bond Yields, Economic Indicators and the Stock Market.

TOOL #36

ECONOMIC INDICATORS: INFLATION

What Is This Tool? Inflation is the general rise in prices of consumer goods and services. The federal government measures inflation with four key indices: Consumer Price Index (CPI), Producer Price Index (PPI), Gross Domestic Product (GDP) Deflator, and Employment Cost Index (ECI)

How Are They Computed? Price indices are designed to measure the rate of inflation of the economy. Various price indices are used to measure living costs, price level changes, and inflation. They are

- Consumer Price Index—the Consumer Price Index (CPI), the most well-known inflation gauge, is used as the cost-of-living index, to which labor contracts and social security are tied. The CPI measures the cost of buying a fixed bundle of goods (some 400 consumer goods and services), representative of the purchase of the typical working-class urban family. The fixed basket is divided into the following categories: food and beverages, housing, apparel, transportation, medical care, entertainment, and other. Generally referred to as a "cost-of-living index," it is published by the Bureau of Labor Statistics of the U.S. Department of Labor. The CPI is widely used for escalation clauses. The base year for the CPI index was 1982 to 1984 at which time it was assigned 100. (See Figure 44.)
- Producer Price Index—similar to the CPI, the PPI is a measure of the cost of a given basket of goods priced in wholesale markets, including raw materials, semifinished goods, and finished goods at the early stage of the distribution system. The PPI is published monthly by the Bureau of Labor Statistics of the Department of Commerce. The PPI signals changes in the general price level, or the CPI, some time before they actually materialize. (Because the PPI does not include services, caution should be exercised when the principal cause of inflation is service prices). For this reason, the PPI and especially some of its subindexes, such as the index of sensitive materials, serve as one of the leading indicators that are closely watched by policy makers. It is the one that signals changes in the general price level, or the CPI, some time before they actually materialize. (See Figure 45.)
- GDP Deflator—the index of inflation is used to separate price changes in GDP calculations from real changes in economic activity. The deflator is a weighted average of the price indexes used to deflate GDP so true economic growth can be separated from inflationary growth. Thus, it reflects price

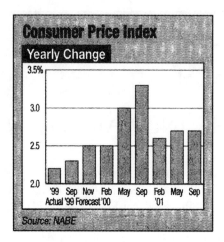

FIGURE 44   How a newspaper charts the CPI. (From *Investor's Business Daily,* September 12, 2000, p. A10. With permission.)

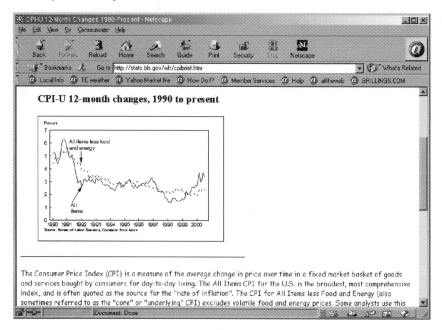

FIGURE 45 How a government web site tracks the CPI.

changes for goods and services bought by consumers, businesses, and governments. Because it covers a broader group of goods and services than the CPI and PPI, the GDP deflator is a very widely used price index that is frequently used to measure inflation. The GDP deflator, unlike the CPI and PPI, is available only quarterly—not monthly. It is also published by the U.S. Department of Commerce.

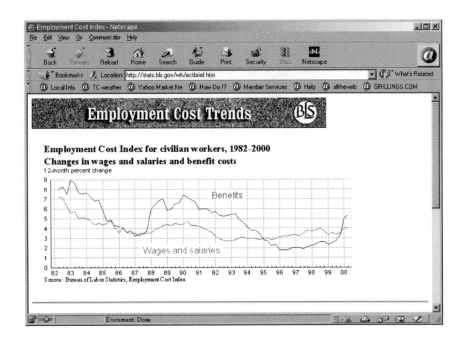

FIGURE 46   How the government charts the ECI at stats.bls.gov web site.

- Employment Cost Index—this is the most comprehensive and refined measure of underlying trends in employee compensation as a cost of production
and measures the cost of labor including changes in wages and salaries and
employer costs for employee benefits. ECI tracks wages and bonuses, sick
and vacation pay plus benefits such as insurance, pension, and Social
Security, and unemployment taxes from a survey of 18,300 occupations at
4,500 sample establishments in private industry and 4,200 occupations
within about 800 state and local governments. (See Figure 46.)

Where Is It Found?   Price indices get major coverage and appear in daily
newspapers and business dailies, on business TV programs like *CNNfn* and *CNBC,*
and on Internet financial news services. Government Internet Web sites
www.stats.bls.gov and www.census.gov/econ/www/ also provide this data.

How Is It Used for Investment Decisions?   Check to see whether the inflation
rate has been rising—a negative, or bearish, sign for stock and bond investors—or
falling, which is bullish.

Rising prices is public enemy No. 1 for stocks and bonds. Inflation usually hurts
stock prices because higher consumer prices lessen the value of future corporate earnings, which make shares of those companies less appealing to investors. By contrast,
when prices rocket ahead, investors often flock to long-term inflation hedges such as
real estate.

See how a chain of events lead from lower rates of inflation to increased consumer spending and possibly, a rising stock market.

Inflation is down *so* real personal income is up *so* consumer confidence jumps *so* consumer spending is up *so* retail sales surge *as* housing starts rise *as* auto sales jump *so* the stock market goes up.

Do note that Federal Reserve Chairman Alan Greenspan is a big fan of the ECI as a good measure to see if wage pressures are sparking inflation.

A Word of Caution: Of course, if inflation disappears, that is no good in the long run, too. Deflation, that is, sharp falling prices is a disastrous event. Think of Texas real estate in the 1980s or California's property woes of the early 1990s. A broader example is the Great Depression of the 1930s.

When demand for goods is so weak that merchants have to brutally slash prices just to stay in business, that is deflation. It leads to layoffs and recession. That is bad for stock investors as profits shrink, but it is good for bond holders—as long as they own a bond backed by an issuer who can pay it back.

Also See: Commodity Research Bureau Indexes, Economic Indicators and Bond Yields, Economic Indicators and the Stock Market, Economic Indicators: Productivity.

Tool #37

Economic Indicators: Money Supply

What Is This Tool?   This is the level of funds available at a given time for conducting transactions in an economy, as reported by the Federal Reserve. The Federal Reserve System can influence money supply through its monetary policy measures. There are several definitions of the money supply: M1 (which is currency in circulation, demand deposits, traveler's checks, and those in interest-bearing accounts), M2 (the most widely followed measure, it equals M1 plus savings deposits, money market deposit accounts, and money market funds), and M3 (which is M2 plus large CDs).

How Is It Computed?   The Federal Reserve System computes these measures.

Where Is It Found?   The weekly money supply figures are released on Thursday afternoons by the Federal Reserve Board and reported in daily newspapers and *The Wall Street Journal* and *Barron's*. (See Figure 47.)

How Is It Used and Applied?   A rapid growth is viewed as inflationary. In contrast, a sharp drop in the money supply is considered to be recessionary. Moderate

MONEY SUPPLY

Money Supply(Bil.$ sa) Latest		Prev.	Yr. Ago
Week ended Aug 14:			
M1 (seas. adjusted)	1090.8	r1088.9	1101.4
M1 (not adjusted)	1084.8	r1088.4	1096.5
M2 (seas. adjusted)	4806.6	r4810.8	4557.6
M2 (not adjusted)	4809.9	r4818.6	4568.4
M3 (seas. adjusted)	6831.9	r6824.5	6203.0
M3 (not adjusted)	6820.8	r6814.8	6200.8

Monthly Money Supply Latest		Prev.	Yr. Ago
Month ended July:			
M1 (seas. adjusted)	1104.1	r1103.2	1101.1
M2 (seas. adjusted)	4792.4	r4779.0	4522.2
M3 (seas. adjusted)	6785.1	r6731.8	6157.2

FIGURE 47   How a newspaper covers the money supply. (From *Barron's Market Week*, August 28, 2000, p. F47. With permission.)

growth is thought to have a positive impact on the economy. Economists attempt to compare with targets proposed by the Fed.

How Is It Used for Investment Decisions? The Fed affects money supply through its monetary policy such as open market operations. The following list summarizes its possible impact on the economy and the stock market.

- Easy Money Policy—the Fed buys securities *so* bank reserves rise *so* bank lending is up *so* money supply is up *so* interest rates are down *as* bond prices rise *so* loan demand goes up *so* the stock market rises.
- Tight Money Policy—the Fed sells securities *so* bank reserves fall *so* bank lending is down *so* money supply is down *so* interest rates are up *as* bond prices fall *so* loan demand is down *so* the stock market falls.

A Word of Caution: A rapid growth (excessively easy monetary policy) is viewed as inflationary and could impact adversely the economy. In contrast, a sharp drop in the money supply is considered to be recessionary and can hurt the economy and the stock market. Moderate growth is thought to have a positive impact on the economy.

Also See: Economic Indicators and Bond Yields, Economic Indicators, Interest Rates, Economic Indicators and the Stock Market.

Tool #38

Economic Indicators: Personal Income and Confidence Indices

What Are These Tools? Personal income shows the before-tax income received by individuals and unincorporated businesses such as wages and salaries, rents, and interest and dividends, and other payments such as unemployment and Social Security.

There are two popular indices that track the level of consumer confidence: one is the Conference Board of New York, an industry-sponsored, nonprofit economic research institute and the other is the University of Michigan's index.

The Consumer Confidence Index measures consumer optimism and pessimism about general business conditions, jobs, and total family income.

The University of Michigan Survey Research Center is another research organization that compiles its own index called the Index of Consumer Sentiment. It measures consumers' personal financial circumstances and their outlook for the future.

How Are They Computed? Personal income data are released monthly by the Commerce Department. The Conference Board's index is derived from a survey of 5000 households nationwide, covering questions that range from home-buying plans to the outlook for jobs, both presently and during the next six months.

The University of Michigan's index is compiled through a telephone survey of 500 households.

Where Are They Found? Daily newspapers, financial television, and online business news services such as America Online and finance.yahoo.com cover these releases. (See Figures 48 and 49.)

How Are They Applied? Personal income represents consumers' spending power. When personal income rises, it usually means that consumers will increase their purchases which will, in turn, affect favorably the investment climate.

The Conference Board's index is considered a useful economic barometer because it provides insight into consumer spending, which is critical to any sustainable economic upswing. Many economists pay close attention to the index, which provides insight into consumer attitudes toward spending and borrowing. Consumers account for two-thirds of the nation's economic activity (i.e., national gross domestic product) and, thus, drive recovery and expansion.

How Are They Used for Investment Decisions? A low or decreased level of consumer confidence indicates concern about their employment prospects and their

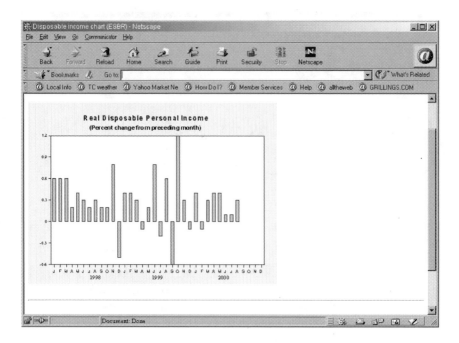

FIGURE 48   How a government web site charts personal income statistics.

New home sales surging

The Commerce Department reported Tuesday that sales of new single-family homes surged at a seasonally adjusted annual rate of 944,000, the highest level in four months.

The 14.7 percent increase was the largest since a 16.4 percent jump in April 1993.

In other news, consumer confidence in the U.S. economy slipped in August, Conference Board figures showed.

The New York-based research group said its index of consumer confidence fell to 141.1 in August from 143 in July, originally reported at 141.7.

Consumer confidence
Index from a survey of 5,000 U.S. households 1985 = 100. Seasonally adjusted
141.1
150
130
110
90
70
M A M J J A
2000

FIGURE 49   How a newspaper covers consumer confidence reports. (From *Sun Sentinel*, August 30, 2000, p. D1. With permission.)

earnings in the months ahead. Uncertainty requires caution in investing. On the other hand, an increased level of consumer confidence spells economic recovery and expansion, thus presenting an investment opportunity. In summary, an increase in personal income, coupled with substantial consumer confidence, is bullish for the economy and the stock market.

A Word of Caution: One must look carefully at how consumers are staying confident. The personal income figures, when measured against spending and borrowing patterns, may show that consumer are dipping into savings or even running up big debts to pay for buying sprees. That is an expansion that is rarely sustainable. To formulate the future prospects about the economy, investors must weigh various economic indicators such as inflation measures.

Also See: Economic Indicators and Bond Yields, Economic Indicators and the Stock Market.

TOOL #39

ECONOMIC INDICATORS: PRODUCTIVITY

What Are These Tools?   Productivity measures the relationship between real output and the labor time involved in its production, or output per hour of work.

How Are They Computed?   The Labor Department compiles productivity figures from its own job surveys that produce unemployment reports and the Commerce Department's work that creates gross domestic product figures. Only business sector output—GDP minus government and not-for-profit organizations—is used in the productivity calculation.

Productivity measures reflect the joint effects of many influences, including changes in technology; capital investment; level of output; utilization of capacity, energy, and materials; the organization of production; managerial skill; and the characteristics and effort of the work force.

Where Are They Found?   Daily newspapers, financial television, and online business news services such as America Online cover these releases. The productivity report is provided quarterly by the Bureau of Labor Statistics of the U.S. Department of Labor. The data are published in a press release and in BLS journals, and computer users can visit www.stats.bls.gov on the Internet for this data.

How Are They Used for Investment Decisions?   Economists consider productivity the key to prosperity. Sizable gains mean companies can pay workers more, hold the line on prices, and still earn the kind of profits that keep stock prices rising. Increased productivity, or getting more worker output per hour on the job, is considered vital to increasing the nation's standard of living without inflation.

A Word of Caution: The productivity statistics mainly cover the manufacturing sector of the economy and do not deal substantively with the large service sector.

Also, note that high productivity may be good for stocks but low productivity is just a mixed bag. Consider that productivity grew at a brisk 2.9 percent annual rate in the 1960s and early 1970s, relatively good times for stock investors. Productivity then slowed to a paltry 1 percent from 1974 through 1995, both horribly weak and extremely strong stock periods. From 1995 through 1998, good times for stocks, productivity grew at around a 2 percent rate, seemingly a new era of productivity driven by computers and other high-tech innovations.

Also See: Economic Indicators and Bond Yields, Economic Indicators: Inflation, Economic Indicators and the Stock Market.

Tool #40

ECONOMIC INDICATOR: RECESSION

What Is This Tool?   Recession means a sinking economy. Unfortunately, there is no consensus definition and measure of recession. In general, it means that the economy is shrinking in size and the number of jobs being lost outnumbers jobs being created.

How Is It Computed?   Here are three primary ways economists define a recession.

1. Three or more straight monthly drops of the Index of Leading Economic Indicators are generally considered a sign of recession.
2. Two consecutive quarterly drops of Gross Domestic Product (GDP) signals a recession.
3. Consecutive monthly drops of durable goods orders, which most likely results in less production and increasing layoffs in the factory sector.

Where Is It Found?   Newspapers, TV shows, and online services all try to guess when recessions start—and end.

How Is It Used for Investment Decisions?   Recession tends to dampen the spirits of consumers and investors and, thus, depress prices of various investment vehicles including securities and real estate.

A Word of Caution: Not all industries in the economy during recession go bad. Some industrial sectors (for example, consumer products industry) are recession resistant or defensive. Investors need to analyze industry by industry.

Also, remember that recessions (or depressions) have political impact for the nation. Political analysts say that President Bush lost the 1992 presidential election—just a year after his leadership of victorious Allied forces in the Persian Gulf War—because the U.S. economy grudgingly recovered from a recession throughout the campaign.

In fact, there is so much political power in defining when the country is actually in recession, that a nonpartisan group of economists known as the National Bureau of Economic Research are the official arbiters of a recession's start and end.

Unfortunately for investors, they act so slowly that it is of little use. Unfortunately for Bush and his recession, the bureau announced one month after Bill Clinton's victory in November 1992 that the nation's ninth post World War II recession officially began in July 1990 and ended in March 1991.

Also See: Economic Indicators and Bond Yields, Economic Indicator, GDP, Economic Indicator: the Index of Leading Indicators, Economic Indicators and the Stock Market, Economic Indicators: Unemployment Rate, Initial Jobless Claims, and Help-Wanted Index.

TOOL #41

ECONOMIC INDICATOR: RETAIL SALES

What Is This Tool?   This figure is the estimate of total sales at the retail level. It includes everything from bags of groceries to durable goods such as automobiles. It is used as a measure of future economic conditions. A long slowdown in sales could spell cuts in production.

How Is It Computed?   Data is issued monthly by the Commerce Department which conducts a mail survey of about 4100 merchants. The previous month's sales figure is an estimate of sales activity based on percentage changes by industry aggregates from older, revised, and more reliable data that is derived from larger samplings. The median revision is a change of 0.2 of a percentage point in sales.

Where Is It Found?   Daily newspapers, financial television, and online business news services such as America Online cover these releases. Commerce Department statistics can be found at www.census.gov/cgn-bin/briefroom/briefrm/ on the Internet.

How Is It Used for Investment Decisions?   Retail sales are a major concern of analysts because they represent about one-half of overall consumer spending. Consumer spending, in turn, accounts for about two-thirds of the nation's Gross Domestic Product (GDP). The amount of retail sales depends heavily on consumer confidence about the economy.

A Word of Caution: This number is volatile and subject to occasionally steep revisions. Remember, too, that strong retail sales could spurt fears of inflation. It could hurt stock and bond markets.

Also See: Economic Indicator: GDP, Economics Indicators: Personal Income and Confidence Indices.

Tool #42

Economic Indicators: Unemployment Rate, Initial Jobless Claims, and Help-Wanted Index

What Are These Tools? Unemployment is the shortage of jobs for people able and willing to work at the prevailing wage rate. It is an important measure of economic health, because full employment is generally construed as a desired goal. When the various economic indicators are mixed, many analysts look to the unemployment rate as being the most important.

Weekly initial claims for unemployment benefits are another closely watched indicator along with the unemployment rate to judge the jobless situation in the economy.

The help-wanted advertising index tracks employers' advertisements for job openings in the classified section of newspapers in 50 or so labor market areas. The index represents job vacancies resulting from turnover in exiting positions such as workers changing jobs or retiring and from the creation of new jobs. The help-wanted figures are seasonally adjusted.

How Are They Computed? The unemployment rate is the number of unemployed workers divided by total employed and unemployed who constitute the labor force. Both statistics are released by the Department of Labor. The help-wanted advertising figures are obtained from classified advertisements in newspapers in 51 major labor markets and an index is compiled by The Conference Board.

Where Are They Found? They are frequently reported in daily newspapers, business dailies, business TV shows, and through online services. Labor Department releases can be found at www.stats.bls.gov for Internet users while Conference Board data is at www.tcb-indicators.org.

How Are They Used for Investment Decisions? An increase in employment, a decrease in initial jobless claims, and a decrease in unemployment are favorable for the economy and the stock market; the opposite situation is unfavorable. The help-wanted index is inversely related to *unemployment*. When help-wanted advertisements increase, unemployment declines, while a decline in help-wanted advertisements is accompanied by a rise in unemployment.

A Word of Caution: No one economic indicator is able to point to the direction to which an economy is heading. It is common that many indicators give mixed signals regarding, for example, the possibility of a recession.

Perhaps the best example of economic theory being turned on its head is the low unemployment figures in 1998 not creating inflationary pressures. Investors, and shoppers, can thank increased productivity and cheap foreign goods for that change.

Also See: Economic Indicators and Bond Yields, Economic Indicators and the Stock Market, Economic Indicator: Recession.

Tool #43

Economic Indicators: U.S. Balance of Payments and the Value of the Dollar

What Are These Tools? A balance of payments is a systematic record of a country's receipts from, or payments to, other countries.

In a way, it is like the balance sheets for businesses, only on a national level. The references you see in the media to the "balance of trade" usually refer to goods within the goods and services category of the current account. It also known as merchandise or "visible" trade because it consists of tangibles like foodstuffs, manufactured goods, and raw materials. "Services," the other part of the category, is known as "invisible" trade and consists of intangibles such as interest or dividends, technology transfers, services (like insurance, transportation, financial), and so forth.

When the net result of both the current account and the capital account yields more credits than debits, the country is said to have a surplus in its balance of payments. When there are more debits than credits, the country has a deficit in the balance of payments.

When deficits in the balance of payments persist, this generally depresses the value of the dollar and can boost inflation. The reason is a weak dollar makes foreign goods relatively expensive, often allowing U.S. makers of similar products to raise prices as well.

How Are They Computed? Trade data is collected by the U.S. Customs Service. Figures are reported in seasonally adjusted volumes and dollar amounts. It is the only nonsurvey, nonjudgmental report produced by the Department of Commerce.

Foreign exchange rates are compiled from trading activity both in bulk transactions among dealers and in commodity markets trading forward contracts.

Where Are They Found? Trade figures and foreign exchange rates are quoted daily in business dailies as well as major newspapers, on computer services such as America Online, and on financial TV networks and specialty shows. (See Figure 50.)

How Are They Used for Investment Decisions? It is necessary for an investor to know the condition of a country's balance of payments, because resulting inflation will affect the market.

Foreign Exchange Rates

Currency	Currency in U.S.$ 9/18	9/15	U.S.$ in Currency 9/18	9/15
Prices as of 3:00 p.m. Eastern Time from Bridge Telerate Markets and other sources.				
f-Argentina (Peso)	1.0007	1.0002	.9993	.9998
Australia (Dollar)	.5423	.5465	1.8440	1.8298
Austria (Schilling)	.0821	.0628	18.110	15.914
c-Belgium (Franc)	.0212	.0213	47.25	47.03
Brazil (Real)	.5408	.5441	1.8490	1.8380
Britain (Pound)	1.4061	1.4000	.7112	.7143
30-day fwd	1.4031	1.3996	.7127	.7145
60-day fwd	1.4017	1.3982	.7134	.7152
90-day fwd	1.4006	1.3971	.7140	.7158
Canada (Dollar)	.6720	.6741	1.4882	1.4834
30-day fwd	.6708	.6735	1.4907	1.4848
60-day fwd	.6703	.6729	1.4916	1.4860
90-day fwd	.6699	.6725	1.4928	1.4869
y-Chile (Peso)	.001774	.001785	563.75	560.25
China (Yuan)	.1208	.1208	8.2771	8.2773
Colombia (Peso)	.000452	.000452	2212.50	2211.50
c-CzechRep (Koruna)	.0242	.0244	41.40	40.98
Denmark (Krone)	.1142	.1153	8.7573	8.6730
Dominican (Peso)	.0625	.0625	16.00	16.00
z-Ecuador (Sucre)	.000040	.000040	25000.00	25000.00
d-Egypt (Pound)	.2809	.2843	3.5600	3.5175
Euro (Euro)	.85460	.85630	1.1701	1.1678
30-day fwd	.85760	.86180	1.1660	1.1604
90-day fwd	.86090	.86520	1.1616	1.1558

Currency	Currency in U.S.$ 9/18	9/15	U.S.$ in Currency 9/18	9/15
Finland (Mark)	.1437	.1454	6.9610	6.8765
France (Franc)	.1305	.1306	7.5756	7.6599
Germany (Mark)	.4369	.4378	2.2886	2.2839
Greece (Drachma)	.002520	.002535	396.85	394.55
Hong Kong (Dollar)	.1283	.1283	7.7965	7.7970
Hungary (Forint)	.0033	.0033	306.57	302.93
y-India (Rupee)	.0218	.0219	45.960	45.750
Indonesia (Rupiah)	.000115	.000115	8670.00	8675.00
Ireland (Punt)	1.0846	1.0979	.9220	.9108
Israel (Shekel)	.2473	.2479	4.0430	4.0345
Italy (Lira)	.000441	.000442	2265.70	2261.07
Japan (Yen)	.009362	.009315	106.93	107.35
30-day fwd	.009391	.009362	106.49	106.81
60-day fwd	.009429	.009400	106.06	106.38
90-day fwd	.009475	.009446	105.54	105.86
Jordan (Dinar)	1.4065	1.4065	.71098	.71098
Lebanon (Pound)	.000661	.000661	1514.00	1513.25
Malaysia (Ringgit)	.2632	.2632	3.7996	3.7995
z-Mexico (Peso)	.105932	.106440	9.4400	9.3950
Netherland (Guilder)	.3874	.3893	2.5812	2.5690
N. Zealand (Dollar)	.4092	.4175	2.4438	2.3952
Norway (Krone)	.1067	.1074	9.3710	9.3105
Pakistan (Rupee)	.0183	.0183	54.75	54.65
y-Peru (New Sol)	.1338	.1301	7.475	7.685
z-Philippines (Peso)	.0218	.0220	45.80	45.50
Poland (Zloty)	.2227	.2222	4.49	4.50
Portugal (Escudo)	.004260	.004313	234.72	231.86

Currency	Currency in U.S.$ 9/18	9/15	U.S.$ in Currency 9/18	9/15
a-Russia (Ruble)	.0361	.0361	27.7300	27.7300
SDR (SDR)	1.28410	1.28650	.7788	.7773
Saudi Arabia (Riyal)	.2666	.2666	3.7504	3.7504
Singapore (Dollar)	.5724	.5734	1.7471	1.7440
SlovakRep (Koruna)	.0199	.0202	50.17	49.55
So. Africa (Rand)	.1374	.1394	7.2800	7.1730
So. Korea (Won)	.000884	.000893	1131.00	1119.80
Spain (Peseta)	.005129	.005190	194.96	192.67
Sweden (Krona)	.1020	.1021	9.8068	9.7910
Switzerland (Franc)	.5609	.5608	1.7828	1.7831
30-day fwd	.5618	.5630	1.7799	1.7762
60-day fwd	.5630	.5642	1.7761	1.7724
90-day fwd	.5644	.5656	1.7718	1.7681
Taiwan (Dollar)	.0320	.0321	31.26	31.15
Thailand (Baht)	.02383	.02391	41.97	41.83
Turkey (Lira)	.000001	.000002	667635	668210
U.A.E. (Dirham)	.2723	.2723	3.6728	3.6728
f-Uruguay (New Peso)	.0807	.0807	12.3980	12.3980
Venezuela (Bolivar)	.0015	.0015	699.0000	688.7500

Special Drawing Rights(SDR) are based on exchange rates for U.S., British, French, German, and Japanese currencies. Source: International Monetary Fund.

Euro: a common currency for 11 European countries. The Federal Reserve Board's index of the value of the dollar against 10 other currencies weighted on the basis of trade was 101.77 Monday , up 0.22 points or 0.22 percent from Friday's 101.55 . A year ago the index was 92.83

a-Russian Central Bank rate, c-commercial rate, d-free market rate, f-financial rate, y-official rate, z-floating rate.

FIGURE 50A   How a newspaper tracks foreign currencies. (From *Investor's Business Daily,* September 19, 2000, p. B22. With permission.)

Currencies abroad	Wednesday (N.Y.)	Today (Tokyo)
Yen per dollar	106.66	106.58
Euro (in dollars)	0.8480	0.8470

Source: Bridge Information Systems

FIGURE 50B   (Copyright 2000, USA TODAY. Reprinted with permission.)

What is better, a strong dollar or a weak dollar? The answer is, unfortunately, it depends. A strong dollar makes Americans' cash go further overseas and reduces import prices—generally good for U.S. consumers and for foreign manufacturers. If the dollar is overvalued, U.S. products are harder to sell abroad and at home, where they compete with low-cost imports. This helps give the United States its huge trade deficit.

A weak dollar can restore competitiveness to American products by making foreign goods comparatively more expensive. Too weak a dollar can spawn inflation, however, first through higher import prices and then through spiraling prices for all goods. Even worse, a falling dollar can drive foreign investors away from U.S. securities, which lose value along with the dollar. A strong dollar can be induced by interest rates. Relatively higher interest rates abroad will attract money dollar-denominated investments which will raise the value of the dollar.

Those Americans owning foreign investments must watch the dollar carefully. A weak dollar makes overseas investments more valuable because assets sold in the foreign currency will yield more dollars. Conversely, a strong dollar will hurt the value of an American's overseas holdings. Assets priced overseas in this scenario would bring home less dollars back from the depressed local currency.

A Word of Caution: Unfortunately, it is difficult to establish a good correlation between the dollar's value and the U.S. stock market's performance. Attention should be focused on the domestic scene as well as international economic developments.

Also See: Economic Indicators and Bond Yields, Economic Indicators and the Stock Market.

Tool #44

The Euro

What Is This Tool? A new currency intended to unite eleven European economies.

How Is It Computed? Well, let us start with a history lesson.

The European Union Treaty (Maastricht Treaty) of 1993 created the European Currency Unit (ECU), a basket of currencies which includes 15 currencies of the European Union (EU) countries. The ECU took two different forms: the official ECU and the private ECU. The official ECU did not trade in the foreign exchange market and was used between central banks and international financial institutions. The private ECU was freely traded in the foreign exchange market, and its exchange rate results from the supply and demand.

Beginning in 1999, as provided in the Maastricht Treaty, the ECU was replaced by the euro. It is based on fixed conversion rates between the currencies of the different countries which will constitute the EMU, thus there is no longer trading in currency rates between ECU members that agreed to use the euro.

According to the Maastricht criteria, the following 11 countries qualified for entry and joined EMU in January 1999: Germany, France, Italy, Spain, Portugal, Belgium, Luxembourg, The Netherlands, Finland, Austria, and Ireland. Of the four remaining EU members, Greece failed to meet the economic conditions for joining and Britain, Sweden, and Denmark opted out for the moment.

Where Are They Found? Foreign exchange rates are quoted daily in business dailies as well as major newspapers, on computer services such as America Online, and on financial TV networks and specialty shows.

How Is It Used for Investment Decisions? That remains to be seen. The introduction of the euro in 1999 is only for operations carried out in the money, foreign exchange, and financial markets. For most retail transactions, the changeover to the euro will only start after the date of the physical introduction of euro banknotes and coins denominated in euros, in 2002 at the latest.

One can assume that investors will watch the euro like they watch any other foreign currency. If European economies are healthy and the financial houses are in order, the euro should perform well. Conversely, weak economics will likely mean a weak euro.

When the euro conversion is complete, perhaps by 2002, investors will notice these changes in Western European investments:

- Stocks will be priced and settled in euro only.
- Government debt will be quoted in euro only.

- Stock and bond deals, such as mergers, will be stated in euro.
- Financial statements from companies and governments will be in euro.

A Word of Caution: This union is an unparalleled event. Any investment which touches Western Europe, whether it be in a security of a European company or agency—or even a U.S. company with large European ties, will be impacted by this three-year conversion. If the euro stumbles, so should investment results of those institutions that it touches.

Tool #45

Footnotes on Newspaper Financial Tables

What Are These Tools?   Owing to the compressed nature of newspaper financial tables, footnotes are used to convey additional information about stocks, bonds, and mutual funds. Footnotes are typically one- or two-letter abbreviations that appear on the line of type reporting a specific stock's results.

Footnotes can add greatly to the understanding of the quotations. Because the Associated Press supplies stock tables to the majority of newspapers, the footnotes tend to be consistent throughout the country.

How Are They Computed?   The Associated Press gets its information directly from the major exchanges and trading houses that supply quotation data.

Where Are They Found?   Most newspaper stock tables include small abbreviations that denote various bits of information about a company, its stock, and its price. The accompanying definitions for those footnotes typically run in a separate box near or with your favorite newspaper's stock, bond, or mutual fund tables.

Here are the translations for the footnote abbreviations used by many newspapers:

u—stock traded at a new 52-week high during the day.
d—stock traded at a new 52-week low during the day.
g—dividend or earnings in U.S. dollars. No yield or P/E unless stated in U.S. money.
n—a new issue in the past 52 weeks. That means the high–low range begins with the start of trading and does not cover the entire 52-week period.
s—split or stock dividend of 25 percent or more in the past 52 weeks. That means that the high–low range is adjusted from the old stock. Dividend begins with the date of split or stock dividend.
v—trading halted on primary market.
x—traded ex-dividend or ex-rights, that is, the first trading day when buyers will not get the previously declared dividend.
y—ex-dividend and sales in full, rather than in 100s as is the rest of the table.
z—sales figures in full, rather than in 100s as is the rest of the table.
pf—preferred issue.
pp—shareholders still owe installments of purchase price.
rt—stock rights.

un—units, typically containing common stock and rights or warrants.
wd—when distributed, shares traded in advance of a stock distribution.
wi—when issued, shares traded in advance of a stock issuance.
wt—warrants.
ww—stock trading with warrants attached.
xw—stock trading without warrants.
vj—company in bankruptcy or receivership or being reorganized or securities assumed by such companies.

The following are some footnotes specifically for dividends. Investors should note that, unless otherwise noted, the listed rates of dividends in stock tables are annual disbursements based on the last quarterly or semiannual declaration.

a—regular dividend with extra dividends.
b—annual rate plus stock dividend.
c—liquidating dividend.
e—declared or paid in preceding 12 months.
i—declared or paid after stock dividend or split up.
j—paid this year, dividend omitted, deferred, or no action taken at last dividend meeting.
k—declared or paid this year, an accumulative issue with dividends in arrears.
r—declared or paid in preceding 12 months plus stock dividend.
t—paid in stock in preceding 12 months, estimated cash value on ex-dividend or ex-distribution date.

These footnotes may appear with NASDAQ listings.

g—dividend or earnings in Canadian money. Stock trades in U.S. dollars. No yield or P/E unless stated in U.S. money.
h—temporary exception to NASDAQ qualifications.

These footnotes may appear with mutual fund tables.

e—ex-capital gains distribution.
s—share dividend or split.
x—ex-cash dividend.
f—previous day's quotation.
nl or n—no front-end load or contingent deferred sales load.
r—redemption fee or contingent deferred sales load may apply.

How Are They Used for Investment Decisions?   Footnotes give an investor information well beyond the typical high–low-and-closing price and trading volume data. The notes can alert an investor to news such as a bankruptcy or an omitted dividend, note the reaching of a peak or trough in share price in a year's time, or briefly help explain a company's dividend history.

A Word of Caution: Footnotes are not always up-to-date so investors are encouraged to double-check before acting on any information contained in these tables. In addition, the Associated Press offers technologies that lets newspapers customize their stock tables; thus, there is limited uniformity of tables.

Also See: Yield on an Investment: Current Yield on a Bond, Current Yield on a Stock, Share–Price Ratios–Price–Earnings Ratio Multiple.

TOOL #46

FT-SE "FOOTSIE" 100 (U.K.) STOCK INDEX

What Is This Tool?   The Financial Times-Stock Exchange 100-Share Index is a capitalized market value index of 100 major companies in the United Kingdom. It is to the British market what the Dow Jones Industrial Average is to American traders.

How Are They Computed?   The Financial Times-Stock Exchange 100-Share Index (known as the "Footsie") is a narrow capitalized-weighted index of the market prices of the 100 most capitalized shares on the exchange. Some companies included in the index are Allied-Domecq, British Airways, British Telecom, Cable & Wireless, Glaxo, Reuters, and Rolls-Royce.

Where Are They Found?   Index information appears in major local and national newspapers and business dailies. Early risers can find live coverage of European markets on financial TV shows. This index can also be tracked by online databases like America Online and on the Internet at web sites like Britain's Money World financial news service at www.moneyworld.co.uk/stocks/ftse100lite/ or at Bloomberg News' www.bloomberg.com.

How Are They Used for Investment Decisions?   The indexes reveal the price performance of United Kingdom stocks. The trend in the indexes should be examined. If prices are unrealistically depressed, a buying opportunity may exist.

These U.K. stock indexes reflect worldwide business conditions and may be used as one indicator for predicting what lies ahead for the New York Stock Exchange. In this manner, it is a technical analysis tool. London is five hours ahead of New York so it also may be used to get a feeling of market sentiment before trading opens on the NYSE. London is the world's third largest stock market. Option contracts may be made in the index. (See Figure 51)

A Word of Caution: Currency fluctuations can play a major part of any overseas investment.

Also See: Dow Jones Industrial Average, Morgan Stanley Eafe Index, German Share Index (Frankfurt Dax).

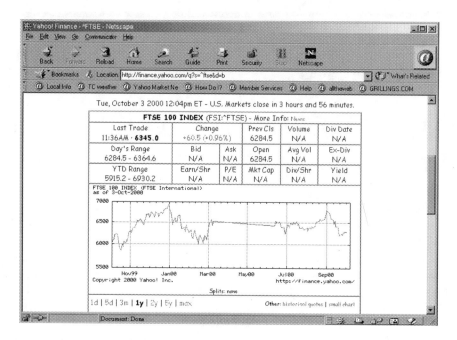

FIGURE 51   Here's how Yahoo reports the Financial Times-Stock Exchange 100-Share Index. (Reproduced with permission of Yahoo! Inc.© 2000 by Yahoo! Inc. YAHOO! and the YAHOO! Logo are trademarks of Yahoo! Inc.)

Tool #47

German Mark

What Is This Tool?   As Germany has been Europe's economic powerhouse, the mark has become one of the world's most important currencies. Its relationship to the U.S. dollar was once a key to the global marketplace and was seen as a barometer of Germany's economic strength vs. that of the United States. Now that Germany has joined the European Union and its currency is linked to the euro, the mark's importance is dwindling.

How Is It Computed?   Marks will be used as the German currency until 2002. So, while its relationship to other currencies will be fixed as a ratio to the new euro currency, it will still be quoted in newspapers and financial reports on television and radio in terms of its relationship to the U.S. dollar. If the mark is at 60, that means that each mark is worth 60 U.S. cents. To figure out what $1 equals in German currency, investors use the following formula.

Example: If 1 mark buys 60 cents, then $1 is equal to 1 divided by 0.60, or 1.66 marks.

Where Is It Found?   Currency rates are listed daily in most major metropolitan newspapers as well as national publications like *The New York Times, USA Today,* and *The Wall Street Journal,* and on computer services such as America Online.

How Is It Used for Investment Decisions?   For American investors buying German securities, the mark's movement is a key part of the profit potential of the investment.

If the mark rises after an investment in German securities is made, the value of those stocks or bonds to a U.S. investor will get a boost from the currency. That is because when the investment is sold, the stronger mark will generate more dollars when the proceeds are converted to U.S. currency. Conversely, a weak mark will be a negative to a German investment for a U.S. investor.

In some cases, the movement of the mark also can be viewed as an indicator of German economic health. This will greatly change in coming years as the new euro takes hold in Europe and its investors' psyche. Still, a strong mark can signal a buoyant European economy, a possible indication to buy German stocks. The mark's strength, however, must be verified not only against the U.S. dollar but against other major currencies outside of the euro circle.

A Word of Caution: Widely quoted currency rates are typically for transactions of $1 million or more. Consumers looking to use such figures to determine currency rates for foreign travel should expect to get somewhat less favorable exchange rates.

Also See: British Pound, Euro, Japanese Yen.

Tool #48

German Share Index (Frankfurt DAX)

What Is This Tool?   This is an index of the market prices of shares of German companies traded on the Frankfurt Stock Exchange. It is Germany's version of the Dow Jones Industrial Average.

How Is It Computed? It is a narrowly determined capitalization-weighted index of 30 of the most active issues. The index is updated continuously on a daily basis. It comprises about 75 percent of German equity value. Some major stocks listed are Bayer, Deutsche Bank, Daimler Chrysler, Volkswagen, and Schering.

Where Is It Found?   Index information appears in major local and national newspapers and business dailies. Early risers can find live coverage of European markets on financial TV shows. This index can also be tracked by online databases like America Online and quote.yahoo.com. (See Figure 52.)

How Is It Used for Investment Decisions?   Index information can be used to determine if German shares are overpriced or underpriced. If can also give an indication of how the Germany economy is faring, which may be useful to those who want to play German bonds or other investments. The index, when compared to individual German shares, may also help determine if specific securities are undervalued and then may be purchased for the capital gain potential.

A Word of Caution: Currency fluctuations can play a major part of any overseas investment.

Also See: Dow Jones Industrial Average, Morgan Stanley Eafe Index.

FOREIGN STOCKS

MONDAY, AUGUST 28, 2000

Stock	Price	Chg
ARGENTINA (Arg. Pesos)		
Banco Frances	7.08	−0.09
Ban Galicia BA	2.99	−0.06
Banco Rio de la	4.15	+0.05
City Equity	3.70	+0.10
PC Holdings SA	1.66	−0.01
Renault Argentin	0.61	−0.01
Siderar S.A.	2.92	...
Siderca	2.17	−0.01
Telecom Argen	4.43	−0.09
Telefon Argen	3.15	+0.04
Electrobas SA	34.78	−1.32
Embratel Partici	40.55	−0.45
Petrobras Pref	54.99	+0.09
Tele Norte Leste	48.15	−0.31
Telesp Celular	25.69	−0.32
Banco de Chile	22.50	...
Banco Santiago	10.30	−0.10
Distribution Y	630.00	−13.00
S.A.C.I. Falabel	540.00	+5.00
AUSTRALIA (Australian $)		
AMP Limited	17.89	−0.35
Austr NZ Bank	13.18	−0.07
Broken Hill Pr	19.60	−0.26
Cable & Wireless	4.38	−0.03
Common BK Aus	29.65	−0.19
Natl Austral Bk	26.10	−0.33
News Corp	22.65	+0.05
News Corp	19.12	−0.08
Telstra Corporat	6.81	+0.09
Westpac Bking	12.87	−0.08
AUSTRIA (Euros)		
Austria Tabakwer	42.70	−1.29
Bank Austria	61.18	+1.08
Erste Bank Der	48.75	+0.25
EVN Energie	35.00	−0.50
EA Generali	184.55	−6.45
Oest Elektriz A	106.30	+0.64
OEMV	85.50	+1.08
VA Stahl	30.00	...
VA Technologie	54.47	−0.23
Wienerb Baus	24.00	−0.07
BELGIUM (Euros)		
AGFA-Gevaert NV	27.19	+0.14
Almanij	48.60	−0.09
Delhaize-LE Lion	63.30	−0.50
Dexia Belgium	159.40	−0.80
Electrabel	247.00	+5.00
AG Fin	35.05	+0.25
Grp Bruxelles La	291.60	−2.20
Kredietbank	52.15	−0.10
Solvay	74.90	−0.10
UCB SA	42.78	−0.12
BRAZIL (Brazilian reals)		
Bradesco	15.75	−0.10
Brahma	1820.01	−58.96
Itaubanco	178.01	+0.02
Petrobras Pref	53.01	−0.49
Telecom de Sao P	32.20	+0.80
BRITAIN Closed for Holiday		
CANADA (Canadian $)		
Bk Nova Scotia	37.55	+0.85
Bk Montreal	62.20	+0.70
BCE	33.55	+0.25
Bombardier Inc	25.15	+0.65
JDS Uniphase Can	184.00	−1.85
Nor Tel	120.75	−1.15
Royal Bk Can	85.10	+0.45
Seagram	84.15	+0.40
Thomson Corporat	55.90	−0.75
Toronto Domin	41.65	+0.85
CHILE (Chilean pesos)		
Empresas CMPC	6251.00	+1.00
COPEC	2290.00	−10.00
Telefon Chile A	2310.00	...
Endesa	196.00	+0.50
Enersis	194.99	−1.01
Entel S.A.	5351.00	+51.00
DENMARK (Danish krones}		
Carlsberg B	312.00	−3.00
D/S 1912 B	103000	+3000.00
Dampskib Sven	142000	+3000.00

Stock	Price	Chg
Danisco	280.00	...
Den Danske Bk	1047.00	−13.00
GN Store Nord	1009.00	+19.00
Nordic Baltic	56.40	−0.40
Novo-Nordisk B	1780.00	+15.00
BG Bank	230.00	...
Tele Danmark	499.00	−6.00
FINLAND (Euros)		
Comptel Plc	18.35	+0.20
HPY Holding A	44.35	+0.10
Fortum	4.00	+0.02
Metso Oyj	13.50	+0.50
Nokia	46.27	−0.03
Sonera Group OYJ	37.13	+1.24
Enso Oy A	10.02	−0.08
Enso Oy A	10.10	−0.25
Tieto Corp	33.30	+0.89
UPM-Kymmene OY	29.12	−0.34
FRANCE (Euros)		
Alcatel Alsthom	86.00	+0.15
AXA	169.6	− 1.4
Carrefour	77.65	+2.05
Elf Aquitaine	226.5	− 3.3
France Telecom	133.5	+ 1.5
L'Oreal	83.6	+1.05
LVMH Moet He	90.65	− 0.8
Sanofi-Synthela	57.5	− 1.2
Total Fr Petr B	171.3	− 1.1
Gen des Eaux	85.6	+ 2.4
GERMANY (Euros)		
Allianz Hldg	389.00	+5.00
Bayer	48.40	−0.26
DaimlerChrysler	61.06	+0.44
Deu Telekom	44.78	+0.19
Deutsche Bk	98.95	−0.84
Veba	56.62	−0.08
Infineon Tech	77.40	+0.60
Muenchener Rueck	324.00	+6.50
SAP AG	276.00	−0.60
Siemens	178.00	+0.50
HONG KONG (Hong Kong $)		
Cheung Kong	98.75	−2.25
China Telecom	60.25	+1.25
Citic Pac	37.70	−2.10
CLP Holdings Ltd	35.60	+0.10
Hang Seng Bk	85.00	−0.25
HSBC Hldgs	108.50	−4.50
Hutchison Wha	108.00	−3.50
Pacific Century	14.80	...
Sun Hung Kai	72.75	−1.25
Johnson Electric	17.55	nt
ITALY (Euros)		
Banca Intesa SpA	4.86	−0.08
Enel SpA	4.53	−0.03
ENI SpA	6.61	+0.10
Generali Assuc	35.38	−0.11
Mediaset	18.79	+0.14
Ist Bn San Paolo	19.48	−0.14
Tecnost SPA	3.69	...
Stet	13.78	+0.03
TIM	9.48	+0.14
Credito Italiano	5.57	−0.01
JAPAN (Japanese yen)		
Mitsubishi Bk	1395	+ 5
Canon Inc.	4930	+ 200
Fujitsu Ltd	3320	− 10
Hitachi	1341	+ 31
Honda	3660	− 50
Matsushita Elec	2975	+ 55
Murata MFG CO.	17050	+ 950
NEC Corporation	3210	+ 110
Nomura Secur	2530	+ 75
Nippon Tele	1270000	+50000
NTT Docomo Inc	2860000	+50000
Rohm Company Ltd	32900	+ 900
7-11 Japan	7020	+ 80
Softbank corp.	14200	+ 210
Sony	11890	+ 790
Sumitomo Bk	1378	+ 3
Takeda chemical	6430	− 60
Tokyo Elec	2455	− 30

FOREIGN STOCK INDEXES

Country & Index	Local Currency			US$	
	Level	Chg	%Chg	YTD %Chg	YTD %Chg
Britain FTSE 100	6563.70	...	...	− 5.29	−14.01
France CAC 40	6615.02	+ 19.91	+ .30	+11.02	− .67
Germany DAXI	7339.22	+ 32.05	+ .44	+ 5.48	− 5.64
Netherlands EOE Index	695.60	+ 4.25	+ .61	+ 3.60	− 7.31
Spain IBEX 35	10907.50	+ 89.20	+ .82	− 6.30	−16.18
Italy MIB Telematico	32623.00	+ 72.00	+ .22	+12.59	+ .73
Switzerland SMI	8338.30	+ 26.60	+ .32	+10.15	+ 2.22
South Africa JSE Indus	8951.10	...	...	− 2.83	−13.90
Canada TSE 300	11224.45	− 21.59	− .19	+33.41	+30.14
Mexico Bolsa	6226.37	+ 45.22	+ .73	−12.67	− 9.97
Argentina Merval	475.31	+ .07	+ .01	−13.65	−13.64
Chile Selective	97.99	+ .07	+ .07	− 2.01	− 4.86
Brazil Ibovespa	17460.33	−182.34	− 1.03	+ 2.16	+ .34
Australia All-Ord	3322.70	− 3.60	− .11	+ 6.88	− 6.84
Japan Nikkei 225	17181.12	+269.79	+ 1.60	− 9.26	−12.68
Hong Kong Hang Seng	17019.76	−216.96	− 1.26	+ .34	...
Singapore Straits Times	2176.11	+ 9.82	+ .45	−12.24	−14.91
New Zealand Top 40	2093.74	...	...	− 5.12	−20.93
Taiwan Taiwan Stock Mkt	7845.87	−180.45	− 2.25	− 7.14	− 6.15
Korea Composite	731.81	+ 2.01	+ .28	−28.82	−27.02

Stock	Price	Chg
Toshiba	1128	+ 4
Toyota Motor	4670	+ 50
MEXICO (Mexican pesos)		
Grupo Financieri	43.10	+0.55
Carsa Global Tel	21.40	+0.50
Cemex SA	42.00	+0.10
Fomento Econ	39.10	...
Grupo Carso	29.15	+0.15
Grupo Financieri	31.50	−0.40
Grupo Televisa	28.20	+0.05
Savia SA	42.85	−0.10
Telef Mexico L	22.95	+0.10
Walmart De Mexic	22.00	+0.35
NETHERLANDS (Euros)		
ABN Amro Hldg	28.25	−0.27
Aegon	45.07	+0.82
Ahold	32.29	−0.01
ASM Lithography	43.97	+0.37
Heineken NV	57.15	−0.50
Intl Nederland	75.13	−0.07
Royal PTT	32.20	+0.53
Philips	54.95	+1.10
Royal Dutch Pe	68.83	+0.41
Unilever Cert	54.90	+0.45
NEW ZEALAND (N.Z. $)		
Auckland Interna	2.95	+0.01
Carter Holt ord	1.80	+0.02
Contact Energy L	2.51	−0.02
Fletcher Energy	8.35	+0.33
Indep Newspap	3.78	+0.01
Natural Gas Corp	1.56	−0.02
Sky Network Tele	3.80	+0.02
Telecom NZ	6.82	+0.40
UnitedNetworks L	6.32	−0.68
Warehouse Group	5.70	nt
NORWAY (Norwegian krones)		
Christiania Bank	47.70	−0.30
Den Norske Bank	39.50	...
Netcom ASA	466.00	...
Norsk Hydro A/S	391.00	+4.00
Nycomed	81.00	−1.00
Opticom ASA	1700.00	−93.00
Orkla A	161.00	...
Petri Geo	164.00	−4.00
Uni-Storebrand	A 65.00	−1.50
Tomra Systems A/	272.00	−6.00
SINGAPORE (Singapore $)		
Chartered Semico	14.80	+0.20
DBS Bank Ltd	21.90	+0.10
Oversea-Chin Bk	12.40	+0.10
Oversea Union B	8.60	+0.25
Pacific Century	22.80	−0.30
Singapore Air	16.90	−0.10
Singap Press F	28.60	...
Singapore Tele	2.77	−0.13

Stock	Price	Chg
Singapore Tech E	2.41	−0.01
United Oversea B	14.00	+0.90
SPAIN (Euros)		
Amadeus Global T	11.60	−0.02
Banco Bilbao R	16.97	+0.07
Banco Popular	34.00	...
Ban Santander	12.00	−0.06
Endesa	22.52	−0.21
Gas Natural	18.80	+0.05
Iberdrola	13.19	−0.11
Repsol	22.28	+0.24
Telefon Espana	21.94	+0.54
Terra Networks S	43.95	+1.35
Terra Networks S	43.95	+1.35
SWEDEN (Swedish kronas)		
ABB Ltd	1080.00	+12.00
Zeneca Group	431.00	−1.50
Ericsson B Fr	186.00	−1.00
ForeningsSparban	142.50	+1.50
Hennes & Mauri-B	174.50	+1.50
Nordic Baltic	64.50	+0.50
Securitas AB	208.50	−1.50
SE Banken	117.00	+2.50
Skandia Forsakri	187.50	+0.50
Svens Han A Fr	152.50	+1.00
SWITZERLAND (Swiss francs)		
ABB Ltd	198.50	+1.00
Adecco SA	1322.00	+17.00
CS Holdings B	383.50	+2.50
Nestle R	3725.00	−15.00
Novartis Reg	2684.00	+10.00
Roche Holding AG	15920.0	+60.00
Schw Ruckvr B	3584.00	+4.00
Swisscom AG	520.00	...
UBS AG	254.00	...
Zurich Ver B	926.00	+9.00
VENEZUELA (Ven. bolivars)		
Banco Provincal	495.00	...
Vencemos I	249.00	...
Vencemos I	225.00	−15.00
Elec Caracas	302.00	−1.25
Fondo de Valores	16.00	−1.00
Manufacturas DE	28.00	nt
Mavesa	46.50	−0.50
Mercantil Servic	640.00	...
Sivensa	22.00	+1.00
Telefon Venez	2160.25	+60.25
hol - market closed for holiday.		
nt - Did not trade.		
Source: Bloomberg Financial		
Markets		

FIGURE 52 How a newspaper covers foreign stock markets. (From *The New York Times*, August 28, 2000, p. C10. With permission.)

Tool #49

Gold Spot Price

What Is This Tool?   The gold spot price is an indication of the market price for a troy ounce of gold purchased today. Gold is a major industrial commodity and a key jewelry component. It was once viewed as the only measure of wealth universally recognized around the globe. There is some debate if this is true any longer.

How Is It Computed?   Gold is a global commodity, thus, there is no one market price. Key spot prices watched by traders include morning and afternoon fixings in Hong Kong and London, and the most current month futures contract for gold on the New York Commodity Exchange.

The price is quoted in troy ounces that come 12 to the pound rather than those traditional ounces for food and liquids that are 16 to the pound.

Where Is It Found?   Gold's spot price is listed in commodity roundups in most metropolitan newspapers as well as *Barron's, The New York Times,* and *The Wall Street journal.* Also, it will be discussed on TV business shows and can be found on computerized databases such as America Online. (See Figure 53.)

How Is It Used for Investment Decisions?   Gold usually does the opposite of common stock. As common stock returns move down, returns on gold move up. In other words, gold compensates for a declining stock market. Transaction costs for gold vary with the type of gold, but the higher the quantity purchased, the lower the percentage commission.

The investor can acquire indirect ownership by purchasing shares in a gold mine. However, the prices of shares do not always move in the same manner as the price of the gold itself. Securities of gold mines do enhance portfolio diversification, though, in the same way that the metal itself does.

The investor also can acquire shares of mutual funds that maintain a strong position in gold stocks or gold bullion. Mutual fund investment offers diversification.

Gold futures can be bought on some commodities exchanges. The investor needs to give only about 10 percent in cash of the contract's value to buy. This low margin requirement provides a leveraging opportunity. Commissions typically are less than 1 percent of the contract's value. Gold futures are traded in several U.S. and foreign exchanges.

Gold ownership has several disadvantages, including:

- Storage costs are high.
- High transaction costs are common.
- Dividend revenue is not received on gold.
- Capital gain or loss potential is significant.

GOLD & SILVER PRICES

Handy & Harman	8/24	8/18	Year Ago
Gold, troy ounce	273.65	275.60	253.80
Silver, troy ounce	4.93	4.86	5.13

Base for pricing gold or silver contents of shipments and for making refining settlements.

Coins	Price	Premium $	Premium %
Krugerrand	280.20	2.10	.74
Maple Leaf	288.20	10.10	3.61
Mexican Peso	337.00	1.40	.49
Austria Crown	274.60	2.00	.72
Austria Phil	288.20	10.10	3.61
U.S. Eagles	288.20	10.10	3.61

Premium is the amount over the value of the gold content in the coin.

ScotiaMocatta Bullion spot gold price 278.15

PLATINUM COIN PRICES

Coins	Coin Price Per Ounce	Premium $	Premium %
Australian Koala	595.70	2.58	15.00
Canadian Maple Leaf	595.70	2.58	15.00
Isle of Man Noble	595.70	2.58	15.00
Amer. Eagle Bullion	595.70	2.58	15.00

Premium is the amount over the value of the platinum content in the coin.

Spot platinum price 580.70. Source: Goldline Int'l, Santa Monica, CA. 800-827-4653
www.GoldlineInternational.com

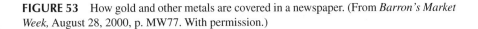

FIGURE 53 How gold and other metals are covered in a newspaper. (From *Barron's Market Week,* August 28, 2000, p. MW77. With permission.)

- Wide price volatility means high risk. It is a speculative investment.
- Certain gold investments are in bearer form. If they are lost or stolen, the owner loses the entire investment. Two examples are bullion and coins.

Gold was once viewed as a defensive investment for times of financial or political upheaval. It was a poor performer in the two decades following its 1980–1981 peak of more than $800 an ounce. It traded at $255 an ounce in late 1999.

Rising gold prices, however, are still viewed as a signal that investor skittishness is rising. Higher gold prices, also, are bullish signals for mining company stocks whose profits are greatly bolstered when the price is up.

A Word of Caution: There are major risks associated with investing in commodity futures and options. These securities provide substantial leverage—that is, the chance to control a huge amount of product for a fraction of the cost. The wrong bet can be financially devastating, however. Option players can only lose their entire investment while futures speculators can be liable for more than they originally invested if they are way off on their hunches.

TOOL #50

GROSS INCOME MULTIPLIER (GIM)

What Is This Tool?   The Gross Income Multiplier (GIM) is a method to compute the price of income-producing property.

How Is It Computed?   The multiplier equals the asking price (or market value) of the property divided by the current gross rental income.

If current gross rental income is $25,000 and the asking price is $300,000, the GIM

$$\$300,000/\$25,000 = 12$$

If similar income-producing properties in the area are selling for "15 times annual gross," this property is undervalued and should be bought. The property would be worth $375,000 (15 × $25,000) in the market.

Where Is It Found?   The GIM for commercial property in an area may be determined by asking real estate brokers and by referring to published real estate data. The investor also should get a feel for the real estate market in the locality by asking around and finding out what similar property has been sold for or is being offered at.

How Is It Used for Investment Decisions?   The GIM approach should be used with caution. Different properties have different operating expenses that must be considered in determining the value of a property.

The GIM is used by the investor to determine an approximate market value of property. A property may be bought if it is undervalued (the multiplier on the property is less than the "going market" multiplier). On the other hand, a property that is overvalued should not be purchased. If the investor currently owns the property and it has a higher multiplier than the "going market" multiplier, the property is overvalued and can be sold before a decline in market price materializes.

Also See: Real Estate Returns: Capitalization Rate (Cap Rate, Income Yield), Real Estate Returns: Net Income Multiplier (NIM).

Tool #51

Herzeld Closed-End Average

What Is This Tool?   The Herzfeld closed end average tracks 20 closed-end mutual funds traded on the stock exchange.

How Is It Computed?   The average is compiled by Thomas J. Herzfeld Advisors and presents the capitalized market value of closed-end funds emphasizing investment in U.S. companies. It equals about 50 percent of the value of all the funds. It is assumed that the same amount is invested in each of the funds with capital gains reinvested in new shares. There is no assumed dividend reinvestment. (See Figure 54.)

Where Is It Found?   It is published in *Barron's*.

How Is It Used for Investment Decisions?   The investor can follow the performance of closed-end mutual funds by examining the trend in the average. An upward trend is a positive sign in a bullish market.

A Word of Caution: Unlike an opened-end fund which constantly sells new shares and its value is based on the underlying securities it owns, the price of a closed-end fund is based on a demand–supply relationship because only its shares trade on a stock exchange. Therefore, the net asset value of the fund may be more or less than its current market value of the securities that the fund owns.

Also See: Lipper Mutual Fund Indexes.

Tracking Closed-End Funds

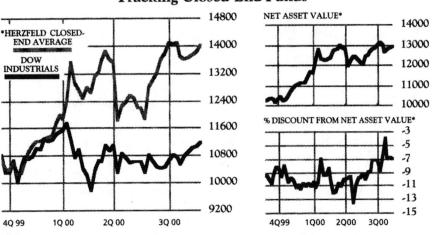

The Herzfeld Closed-End Average measures 15 equally-weighted closed-end funds based in the U.S. that invest principally in American equities. The net asset value is a weighted average of the funds' NAVs. *The net asset value and % discount charts lag the market by one week. Source: Thomas J. Herzfeld Advisors Inc., Miami. 305-271-1900

FIGURE 54   How a newspaper covers closed-end fund performance. (From *Barron's Mutual Funds,* August 28, 2000, p. F72. With permission.)

Tool #52

Index Arbitrage

What Is This Tool?   Index arbitrage is defined as the purchase or sale of a basket of stocks in conjunction with the sale or purchase of a similar derivative product, such as index futures. The goal is to profit from the price difference between the basket and the derivative product.

How Is It Computed?   Index arbitrage profit = $Y_b - X_a$ where

Y_b = Price of higher-priced basket of securities on Exchange B
X_a = Price of lower-priced index of those same securities on Exchange A

(Profits can be made in the reverse order as well, when the stocks are expensive and the related index is cheap.)

Example: An investor buys predetermined baskets of stocks on the floor of the New York Stock Exchange for $5 million and simultaneously sells $5.05 million worth of related index futures contracts on the floor of the Chicago Board of Trade (CBOT), hoping to profit from the price differences on the two exchanges.

$$\text{Arbitrage profit} = \$5,050,000 - \$5,000,000 = \$50,000$$

The arbitrage profit is $50,000 on this transaction. This trade and similar transactions would increase the demand and, therefore, the price of the NYSE securities while simultaneously lowering the price of the CBOT index. This should continue until the prices of the two securities were in parity.

Index arbitrage is most visibly played in a tactic known as program trading. It is defined as a wide range of trading strategies involving the purchase or sale of 15 or more stocks having a total market value exceeding $1 million. In typical market conditions, one-half the traders using program trading use it for arbitrage purposes to profit from the price discrepancies between the stock and so-called derivative markets where stock-like investments trade. The other one-half use program trading for various portfolio tactics.

Where Is It Found?   An investor seeking index arbitrage profits can find what "premium" or "discount" key market index derivatives are trading at vs. the true value of the underlying stocks in various computer online databases. When price differences are noted, they must be taken advantage of instantly.

Program traders do make money. A study that covered 2,659 S&P 500 index arbitrage trades found that program traders produced superior risk-adjusted returns.

However, this is a job for the nimble traders. Over the six-month period that was studied, the profitable situations lasted on average about three minutes.

How Is It Used for Investment Decisions?   Actually, limits on program trading during the trading day signal market volatility.

In response to concerns that index arbitrage may be aggravating large market swings, the New York Stock Exchange instituted a set of program trading restrictions. When the Dow Jones Industrial Average moves 50 points or more from the previous day's close, program trading curbs essentially stop a key computer, so program trading must be done by hand. This so-called "collar" limits the powerful swings program trading can bring to the market.

When index arbitrageurs are being "collared," an investor knows that program trading is having a smaller impact on price moves. That is no guarantee, though, that market volatility will dampen. Because 50 Dow points is a relatively small percentage swing in a day's trading, regulators are considering loosening the "collar" to a much higher point amount that start in 1999.

A Word of Caution: Unfortunately, you will need a big checkbook to play index arbitrage. About $25 million is required, and high transaction costs involved make stock index futures arbitrage unsuitable for most individuals.

Tool #53

Indexing

What Is This Tool? An index is assigned in order to compare a financial statement account or item covering at least three years. In computing a series of index numbers, a base year is selected and all other years are compared to it.

How Is It Computed? A base year that is most representative (typical) of the company's operations is selected and assigned an index of 100. All index numbers are computed by reference to the base year.

$$\text{Index} = \frac{\text{Current year amount}}{\text{Base year amount}}$$

Where Is It Found? The investor should determine index numbers for financial statement items important to him that are found in the company's annual report. The investor may look at the trend in sales, net income, total assets, and so on.

Example: The base year is 20X1 and net income was $6,000,000. Thus, an index of 100 is assigned for 20X1. Net income for 20X2 and 20X3 were $6,600,00 and $5,000,000, respectively. The index numbers are

$$20X2 \ \$6,600,000/\$6,000,000 = 110$$
$$20X3 \ \$5,000,000/\$6,000,000 = 83$$

The sharp decline in net income should be of concern to the investor. A declining profitability spells financial troubles for the company. In such a case, an investment in the company may not be warranted.

How Is It Used for Investment Decisions? When a comparison of accounts covering three years or more is made, the year-to-year method of comparison may become too cumbersome. The best way to look at a long-term trend for comparison purposes is through the use of index numbers. The investor can identify any financial statement accounts or items that appear out of line.

If sales and profitability of the company are significantly increasing, growth is indicated. This may be a time to buy the stock. However, the investor who owns a stock in a company that has a drastically falling revenue base should consider selling it.

This kind of logic is also used to compare investments over similar time periods. For example, instead of looking at, say, total returns of two mutual funds, an investor might look at how each fund did if equal amounts of money were put into the two funds at a starting point. Many fund companies and fund trackers talk of how $10,000 in various funds would have performed over several years.

A Word of Caution: As in the case with the computation of year-to-year percentage changes, certain changes, such as those from negative to positive amounts, cannot be expressed by means of index numbers. Further, the base year selected as most typical may, in retrospect, not be appropriate.

Also See: Profitability: (Trend) Horizontal Analysis.

Tool #54

Initial Public Offerings (IPOs)

What Are These Tools?   Initial public offerings are companies issuing stock to the public for the first *time.*

How Are They Computed?   Lists of companies "going public" through initial public offerings are compiled from information filed at the Securities and Exchange Commission and from brokerages underwriting these new issues. Numerous services, such as Securities Data, tally data on the number of companies going public and the lead underwriter that handled the deals, as well as the value and number of shares sold.

Where Are They Found?   Publications such as *Barron's, The New York Times,* and *The Wall Street Journal* publish weekly listings of recent SEC new-issue filings and issues in the SEC review process expected to be sold in the coming week. Data of new issue volume are reported in *The Wall Street Journal* as part of a quarterly wrap-up of how the major brokerages fared in this lucrative business of taking companies public. News on IPOs can be found at biz.yahoo.com/reports/ipo.html or at www.ipocentral.com. (See Figure 55.)

How Are They Used for Investment Decisions?   Lists of upcoming new issues can provide an investor with three potential insights.

First, a general look at investor appetite for stocks. When demand is high, there are usually many new issues.

Second, it is a chance to learn the name of companies that offer the chance to get in on the ground floor.

Finally, a contrarian view; it can be viewed as a signal of what industries—or equities in general—to shun when initial public offerings soar. A surge in initial public offerings is often viewed as a signal that the market as a whole, or a sector having a flurry of new issues, has peaked in its current cycle.

A Word of Caution: Initial public offerings have great investment sex appeal. However, they are filled with risks. Studies have shown that investors lose money more than one-half the time when buying new shares.

	Ticker Code	Initial Offer Price	IPO Date	Most Recent Price
SELECTED INITIAL PUBLIC OFFERINGS				
ServiceWare Tech	SVCW	7.00	8/25	8.75
O2Micro Intl Ltd	OIIM	9.00	8/23	17.50
Ista Pharmaceutic	ISTA	10.50	8/22	11.94

FIGURE 55   How a newspaper covers initial public offerings. (From *Barron's Market Week*, August 28, 2000, p. MW72. With permission.)

TOOL #55

INSIDER TRADING ACTIVITY

What Is This Tool?   Insiders are corporate directors, officers, other executives of the company, and stockholders who own 5 percent or more of the company's voting shares. Insiders know important information that is not publicly available. Insiders' trading activity refers to the number of shares bought or sold by insiders.

How Is It Computed?   Insiders must disclose their buying and selling activity to the Securities and Exchange Commission. With that information at hand, many kinds of analysis can be made. One is called "net insider share volume," and equals insider shares bought less insider shares sold.

Another analysis is the "buy–sell ratio." Imagine that QQQ Inc. has insiders buying 60,000 share and selling 40,000.

$$\text{Insider buy–sell ratio} = \frac{\text{Insider shares bought}}{\text{Insider shares sold}}$$

$$\text{Insider buy–sell ratio} = \frac{60,000}{40,000}$$

$$= 1.5$$

A higher ratio is more bullish while a lower ratio is more bearish.

Where Is It Found?   Insiders' trading activity can be tracked by watching the SEC's Web site on the Internet at www.sec.gov. Insider information is also available in various newsletters and financial publications and is occasionally covered in *The Wall Street Journal, Barron's, Investor's Daily,* and local newspapers.

How Is It Used for Investment Decisions?   When officers or directors of a company are purchasing significantly more shares than they are selling, this is a bullish sign—perhaps earnings will be growing. On the other hand, if officers and directors are selling their shares rather than buying, something unfavorable may be on the horizon. In bear markets the signs may not be as accurate because the insiders may be selling for tax or personal reasons. Some executives may be selling shares to obtain cash to exercise stock options.

The most knowledgeable investors are company insiders because they are familiar with the current and prospective financial happenings of the business. Insiders typically earn unusually high returns after considering any risks involved. After all, they have information not yet publicly released. If insiders are buying, the investor may do likewise. If insiders are selling, so should the investor.

A Word of Caution: There is a timeliness problem because insider information may be several months old by the time it is released. Therefore, the investor's timing is crucial or else there will be little benefit to knowing insider transactions.

Insiders may be trading the stock for other reasons than inside information to profit on. For example, an officer may sell shares to obtain extra funds to buy a new home or, in fact, may be selling certain shares to cash in a larger option later down the road. Although the insider has sold shares, the price of the stock may, in fact, increase. (See Figure 56.)

INSIDER TRADES

Insider Trades is a weekly report of stock transactions involving officers, directors and owners of 10 percent or more of a publicly held company. The trades were taken from documents filed with the Securities and Exchange Commission and the stock exchanges.

OPEN MARKET PURCHASES

Company (Exchange)	Insider	Title	$ value	No. of shares	% of hold.	Transaction dates
Bentley Phrmctcls (AMEX)	R. Cleveland [1]	D	411,069	52,100	3.00	08/22/00
Catalina Lighting (NYSE)	D. S. Rappaport	VP	87,500	50,000	48.00	08/24/00
Catalina Lighting (NYSE)	R. Hersh	CB	78,750	45,000	25.00	08/24/00
Commercial Net Lease (NYSE)	G. Ralston	P	106,900	10,000	4.00	08/24/00
Consulier Engnrng (NASDAQ)	W. B. Mosler [2]	CB	35,400	20,000	0.5	08/29/00
Elcotel (NASDAQ)	D. S. Fragen	O	33,416	22,300	149.00	07/21-08/29/00
Europa Cruises (NASDAQ)	D. A. Vitale [1]	CB	7,845	25,000	0.5	08/16-21/00
Hollywood.Com (NASDAQ)	L. S. Silvers	P	37,400	5,000	0.4	08/29/00
Hollywood.Com (NASDAQ)	M. Rubenstein	CB	37,400	5,000	0.4	08/29/00
Mayors Jewelers (AMEX)	M. A. Gilliam	D	135,262	40,000	400.00	08/17-23/00
Proxymed (NASDAQ)	M. K. Hoover	CB	191,265	135,000	68.00	08/24-31/00
Transmedia Network (NYSE)	Valuevision Intl [2]	Z	410,364	89,992	70.00	08/23/00

OPEN MARKET SALES

Company (Exchange)	Insider	Title	$ value	No. of shares	% of hold.	Transaction dates
Artesyn Technolgs (NASDAQ)	F. C. Lee	O	273,000	7,000	21.00	08/24-30/00
Artesyn Technlgs (NASDAQ)	R. D. Schmidt [1]	D	3,810,440	100,000	12.00	08/25-29/00
Checker Drive-In (NASDAQ)	R. B. Maggard	D	21,494	4,583	100.00	08/02/00
Dixon Ticonderoga (AMEX)	D. R. Gardner [1]	Z	1,677,270	559,100	100.00	06/28-08/03/00
Ivax (AMEX)	E. R. Biekert	D	47,730	1,000	3.00	07/20-25/00
Noven Pharmctcls (NASDAQ)	S. Becher [1]	D	196,526	6,000	30.00	08/30/00
Noven Pharmctcls (NASDAQ)	S. Sablotsky [1]	CB	34,901,705	1.08M	52.00	08/14-28/00

OTHER ACQUISITIONS

Company (Exchange)	Insider	Title	Activity type	No. of shares	% of hold.	Transaction dates
Mastec (NYSE)	J. S. Sorzano	D	NonOpnMkt	2,500	62.00	05/02/00

OTHER DISPOSITIONS

Company (Exchange)	Insider	Title	Activity type	No. of shares	% of hold.	Transaction dates
Artesyn Technlgs (NASDAQ)	P. A. Oreilly	O	NonOpnMkt	81,721	50.00	08/01/00
Engle Homes (NASDAQ)	D. Shapiro [1]	VP	NonOpnMkt	17,454	6.00	08/07/00

KEY TO TITLES: CB, chairman of the board; P, president; O, officer; D, director; Z, shareholder or affiliated person. OTHER FOOTNOTES: [1] Indirect holdings; [2] Also holds other types of securities; [3] Option-related transaction.
SOURCE: First Call/Thomson Financial, Boston.

FIGURE 56 Here's how a daily newspaper wraps up insider trading activity. (From *Miami Herald,* September 18, 2000, p. 45. With permission.)

Tool #56

Interest Rates: Fed Funds, Discount, and Prime

What Are These Tools?   These are three key interest rates closely tied to the banking system. The discount rate is the rate the Federal Reserve Board charges on loans to banks that belong to the Fed system. The federal or fed funds rate is the rate that bankers charge one another for very short-term loans, although the Fed heavily manages this rate as well. The prime rate is the much discussed benchmark rate that bankers charge customers.

How Are They Computed?   The three rates, at times, work in tandem.

The Fed Funds rate is the major tool that the nation's central bank, the Federal Reserve, has to manage interest rates. Changing the target for the Fed Funds rate is done when the Fed wants to use monetary policy to alter economic patterns.

The discount rate was once the Fed's key tool. Now, it takes a backseat as a largely ceremonial nudge to markets made often after Fed Funds changes are implemented.

The prime rate is a heavily tracked rate although it is not as widely used as a corporate loan benchmark as it has been in the past. The prime is set by bankers to vary loan rates to smaller businesses and on consumers' home equity loans and credit cards.

Where Are They Found?   Changes in the Fed Funds rate target and prime lending rate typically get lots of TV news time and are often front-page news, particularly in times of economic troubles. Many major newspapers also carry a daily rate tally that includes the discount, fed funds, and the prime rates. (See Figure 57.)

How Are They Used for Investment Decisions?   A watcher of these three interest rates will see a reflection of the banking system's view of the strength of the U.S. economy.

These rates tend to rise when the Federal Reserve is in a tight money posture, when it tries to keep an expanding economy from overheating and creating too much inflation. This is often viewed as a negative signal for both stock and bond investments as a cooled economy and higher rates can hurt many securities prices.

Conversely, the rates tend to fall when the Federal Reserve is in an easy money mode, when it wants available credit and low interest rates to stimulate a moribund U.S. economy. For the risk-taking investor, this can be a signal to boost stock and bond holdings. An expected economic recovery may boost equity issues while falling rates should be a boon to bond prices.

Money rates

Data include indexes used by financial institutions to make investments or to set rates. As these indexes move, rates and yields are likely to move.

Selected loan rate benchmarks

Prime rate: 9.50%

Discount rate: 6.00%

Broker call loan rate: 8.25%

Federal funds market rate: High 6.50; Low 6.50; Last 6.50%

Dealers commercial paper: 30-180 days, 6.48-6.51%

Commercial paper by finance company: 30-270 days, 6.47-6.43%

11th District cost of funds: 5.36%; Month ago, 5.20%

IBC's money fund report average: 7-day yield simple, 6.17%

Bond buyers 20-bond index: 5.51%; Month ago, 5.61%

Selected mortgage rate benchmarks

FHLMC U.S. Average commitment: 80% loan-to-value mortgage, 8.04%

FHFB Nat'l Average Contract rate: Previously occupied homes, 8.17%

FNMA posted yields, 30-day lock: (30-year commitment, priced at par) 8.07%

FNMA posted yields, 60-day lock: (30-year commitment, priced at par) 8.12%

The above indexes are used by financial institutions in determining rates and yields on consumer deposit and loan products, as reported by bankrate.com as of Thursday or by Dow Jones National Markets as of Friday

SOURCE: BANKRATE.COM AND DOW JONES NATIONAL MARKETS

FIGURE 57 Here's how a daily newspaper wraps up interest rate trading. (From *Sun Sentinel*, August 21, 2000, p. 58. With permission.)

At times, these rates may send contradictory signals because bankers and the Federal Reserve may disagree on the economy's direction.

Also See: Cash Investment: Money Market Fund Yields, Bond Market Indexes, Economic Indicators and Bond Yields, Interest Rates: Three-Month Treasury Bills.

Tool #57

Interest Rates: 30-Year Treasury Bonds

What Is This Tool?   The most widely watched interest rate in the world, the security known as the "T-bond" is seen as the daily barometer of how the bond market is performing. The 30-year Treasury bond is a fixed-rate direct obligation of the U.S. government. There are no call provisions on Treasury bonds.

How Is It Computed?   Traders watch the price of the U.S. Treasury's most recently issued 30-year bond, often called the "bellwether." The price is decided by a series of dealers who own the exclusive right to make markets on the bonds in U.S. markets. (The bond trades around the clock in foreign markets.) Bond yields are derived from the current trading price and its fixed coupon rate.

Where Is It Found?   The T-bond price and yield can be found in the credit market wrap-up in most newspapers and on the TV business shows plus on computer databases such as America Online. One web site on the Internet from Bloomberg News, www.bloomberg.com, has extensive coverage on the Treasury market.

How Is It Used for Investment Decisions?   Traders who hold T-bonds are exposed to a great deal of risk. Their willingness or unwillingness to take on that risk—and the resulting changes which that brings to 30-year bond prices and yields—is often viewed as a proxy on the long-term outlook for the U.S. economy.

Because of its long-term nature, the T-bond is extra sensitive to inflation that could ravage the buying power of its fixed-rate payouts. Thus, the T-bond market, also, is watched as an indicator of where inflation may be headed.

Also, T-bond rates somewhat impact fixed-rate mortgages. (These loans are usually more tied to 10-year Treasury rate.) Still, the T-bond yield is also seen as a barometer for the housing industry, a key leading indicator for the economy.

A Word of Caution: While 30-year Treasury bonds often offer sexy yields, they are also very volatile investments. A one percentage point move in interest rates can cost a trader 10 percent or more of their principal on such holdings, for example.

In addition, the Treasury only began selling 30-year bonds in 1997. When 30-year yields fell below 5 percent in mid-1998, many news accounts called the rates "record lows." While technically correct, it was historically misleading. Long-term Treasury yields (the government previously sold 20-year bonds) had not been that low since the early 1960s.

Also See: Interest Rates: Three-Month Treasury Bills, Mortgage Rates.

Tool #58

Interest Rates: Three-Month Treasury Bills

What Is This Tool?   The Treasury bill rate is a widely watched rate for secure, cash investments. In turbulent times, the rate can be volatile and can be viewed as a signal of the economy's health.

How Is It Computed?   T-bills, both three-month and six-month issues, are auctioned every Monday by the U.S. Treasury through the Federal Reserve.

Rather than pay interest, the securities are sold at a varying discount defending the prevailing interest rate. The investors get their interest by redeeming the T-bill at full face value when it matures.

The government reports the average effective interest rate it paid each week.

Where Is It Found?   Results of the Monday auctions can be found in most major daily newspapers' business sections such as *The New York Times* and *The Wall Street Journal*. Trading in the secondary markets for T-bills also is reported in these papers as well as in most major daily newspapers and on business TV channels such as CNBC. (See Figure 58.)

How Is It Used for Investment Decisions?   The T-bill rate shows what can be expected to be earned on no-risk investments. Historically, T-bills have returned little more than the inflation rate. Many conservative investors buy T-bills directly from the government. T-bill rates approximate rates on money market mutual funds or statement savings accounts, also popular savings tools for the small investor.

When these low-risk rates are high or rising, it can be a negative signal for stocks and bonds because in such situations individual investors tend to shy away from riskier investments. Ironically, T-bill rates often rise in anticipation of an economic strengthening—a bullish sign for stocks.

Conversely, when T-bill rates are low or falling, small savers tend to look to markets like stocks, real estate, or bonds to beef up their returns. However, experts contend that falling T-bill rates may show economic weakness, which is not a healthy situation for the market.

A Word of Caution: Short-term rates can fluctuate greatly in times of economic uncertainty. Thus, their ability to indicate longer-term trends can be impacted by very short-term events.

Also See: Cash Investments: Certificate of Deposit Yields, Cash Investments: Money Market Fund Yields, Interest Rates: 30-Year Treasury Bonds.

Short-term rates increase

Interest rates on short-term Treasury securities rose in Monday's auction with three-month bills hitting their highest level in nine years.

The Treasury Department sold $9.5 billion in three-month bills at a discount rate of 6.110 percent, up from 6.090 percent last week. An additional $8.5 billion was sold in six-month bills at a rate of 6.090 percent, up from 6.075 percent.

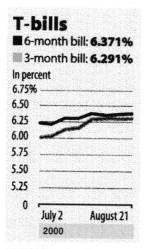

T-bills

■ 6-month bill: **6.371%**
▨ 3-month bill: **6.291%**

In percent

6.75%
6.50
6.25
6.00
5.75
5.50
5.25
0

July 2   August 21
2000

SOURCE: Bloomberg News

The three-month rate was the highest since Jan. 28, 1991, when the bills sold for 6.22 percent. The six-month rate was the highest since July 31 when the rate was 6.115.

The new discount rates understate the actual return to investors — 6.291 percent for three-month bills with a $10,000 bill selling for $9,843.90 and 6.371 percent for a six-month bill selling for $9,692.10.

FIGURE 58   Here's how a T-bill auction is covered by a newspaper. (From *Sun Sentinel*, August 22, 2000, p. D1. With permission.)

TOOL #59

INVESTING STYLES

GROWTH INVESTING

What Is This Tool? This strategy used by many well-known money managers, such as Fidelity Magellan Fund's former boss Peter Lynch and American Century Investors, is associated with stocks of fast-growing companies. "Growth" issues are often considered risky because their share-prices trade at high price–earnings ratios compared to other shares in the market.

How Is It Computed? Growth-oriented portfolios traditionally contain stocks that have higher-than-average sales and growth rates. An investor seeking growth investments must obtain these or other ratios of growth for the investment he is reviewing as well as key market benchmark or peer securities.

Example:

Security	5-Year Sales Growth Rate	5-Year Profit Growth Rate
MSL Technology	45%	33%
Old Line Manufacturing	10%	9%
GeeWhiz Growth Fund	33%*	40%*
S&P 500	18%*	20%*
*Average		

In this example, MSL Technology and the GeeWhiz Growth Fund are the growth selections. The company, and the companies in the fund, are seeing sales and profits grow faster than, for example, the averages for the Standard & Poor's 500-stock index.

Where Is It Found? Many brokerage reports or computerized investment databases can be used to locate a company's key ratios and then compare them to market or industry averages. Another way to find lists of growth stocks is to get copies of semi-annual reports from growth mutual funds. A growth investor "screens" out stocks with below-average sales and/or profit growth. For mutual funds, Morningstar Inc. reports clearly detail average earnings growth ratios for each fund as well as describe what funds have a growth bias.

How Is It Used for Investment Decisions? Growth investing is a strategy that calls for the investor to try to buy the "best" companies, often regardless of share prices. This strategy appeals to many investors because they feel comfortable buying expanding, profitable companies. In fact, most of the best performing equity mutual funds, for example, have practiced growth investing.

Fast-growing companies typically see their share prices grow through a constant price multiple being applied to their expanding profits—or in the best case, an increased multiple given by the market to the growing profits.

A Word of Caution: Growth stocks are considered risky because they tend to trade at above-market price-to-book and price-earnings ratios. This makes them very susceptible to wide price swings when earnings or sales start to slow down. Like many other investment disciplines, growth stocks have their cycles of investor popularity and disfavor. When growth stocks are out of favor, investors tend to buy up "value" or "cyclical" shares that can have below-average price-to-book and/or price-earnings ratios.

Investors must carefully check the sales and earnings of fast-growing companies. Is the increase sustainable? Or is it coming from short-lived events such as the sale of assets or accounting gimmickry? Thus, ratios alone cannot be the only guide to an investment decision. Only fundamentals, such as management's abilities, overall profitability, and industry positioning, should be checked before investing.

Also See: Morningstar Mutual Fund Rankings, Share Price Ratios: Price-Earnings Ratio (Multiple).

VALUE INVESTING

What Is This Tool?   This strategy is associated with buying out-of-favor stocks. "Value" issues often are perceived to have lower risks because their share prices already have been discounted compared to other shares in the market.

How Is It Computed?   Value-styled portfolios traditionally contain stocks that have lower-than-average price-to-book ratios and/or price-earnings ratios. An investor seeking value investments must obtain these or other ratios of value for the investment he or she is reviewing as well as key market benchmark or peer securities.

Example:

Security	Price-to-book ratio	Price-earnings ratio
XYZ Corp.	2.2	19
Widget Growth Fund	3*	32*
Manufacturing Co. of Iowa	0.8	6
S&P 500	1.9*	16*
*Average		

In this example, Manufacturing Company of Iowa's stock is the value selection. Its shares are trading not only at a discount to the other choices to buy, it is also at a discount to averages for the Standard & Poor's 500-stock index.

Where Is It Found?   Many brokerage reports, *Barron's,* and computerized investment databases can be used to locate a company's key ratios and then compare them to market or industry averages. A value investor "screens" out above-average or overpriced issues before making selections. For mutual funds, Morningstar's fund clearly reports detail average price-to-book and price–earnings ratios for each fund as well as describing what funds have a value bias.

How Is It Used for Investment Decisions?   Value investing is a somewhat contrary art where the investor truly tries to buy low and sell high. In theory, the purchase of value stocks at these lower multiples should reduce the risk of major losses.

However, there may be good reasons for a company's stock to be valued lower than its peers'. Factors such as stiff competition, poor management, and heavy debts could be scaring off investors. These are problems that could eventually severely damage the company and create large investment losses.

Value investing has its cycle like other investment disciplines. In the late 1980s, value investors were left way behind the pack when growth-oriented stocks—issues that tend to trade at high multiples—were investor favorites.

A Word of Caution: Like many other investment disciplines, value stocks have their cycles of investor popularity and disfavor. When value stocks are out of favor, investors tend to buy up "growth" shares regardless of their steep, premium price-to-book, and/or price–earnings ratios.

There are also reasons beyond simple ratios that certain stock prices are depressed. In some cases, these below-average figures are tipping off investors that such potential "value" shares are poor investments and that the share price only will go lower.

Thus, ratios alone cannot be the only guide to an investment decision. Only fundamentals, such as management's abilities, overall profitability, and industry positioning should be checked before investing.

Also See: Share–Price Ratio: Price Book Value Ratio, Share–Price Ratios: Price–Earnings Ratio Multiple, Morningstar Mutual Fund Rankings.

Tool #60

Investor's Business Daily's
Smart Select™ Ratings

What Is This Tool?   According to Investor's Business Daily, "SmartSelect™ Corporate Ratings consist of five proprietary evaluations that when used together, provide one of the most powerful stock selection tools available. These fundamental and technical ratings are based on the most extensive study conducted on the characteristics of stock market winners."

How Is It Used for Investment Decisions?   According to Investor's Business Daily, "Companies with high SmartSelect™ Ratings may be prospects for further research. We suggest you look for stocks with the following minimum ratings: Earnings Per Share Rating of 80 or higher; Relative Price Strength Rating of 80 or higher; Industry Group Rating of A; Sales + Profit Margins + Return on Equity Rating a minimum of B; Accumulation/Distribution Rating a minimum of B."

How Is It Computed?   Investor's Business Daily's stock charts include five proprietary rankings for each stock listed:

Earnings Per Share (EPS) Rating EPS Rating is determined by comparing a company's two most recent quarter's earnings growth rate to the same quarters one year prior. Then, the company's three to five year annual growth rate is examined. The results are compared to all other companies in the Investor's Business Daily database and rated on a scale from 1 to 99, with 99 being the highest. An EPS rating of 99 means the stock is in the top 1% in terms of earnings and has outperformed 99% of all other stocks.

Relative Price Strength (RS) Rating Relative Price Strength (RS) Rating measures the overall price performance of a given stock against the rest of the market. RS Rating is calculated by comparing a company's price changes over the past 52 weeks to all other stocks in the market. The stock is then assigned a performance rating from 1 to 99. An RS rating of 99 means the stock has outperformed 99% of all other stocks over the past year.

Industry Group Relative Strength (Grp RS) Rating Industry Group Relative Strength Rating compares a stock's industry group price performance over the past six months to Investor's Business Daily's other 196 industry groups. Within this precise breakdown, a stock's industry is graded on an A to E scale; an A rating representing an industry that performs among the top 20%; an E, on the other hand, identifying a laggard industry within the bottom 20%.

Sales + Profit Margins + ROE (SMR) Rating A quick way to evaluate a company's future earnings potential is to look at the Sales + Profit Margins + ROE (SMR) Rating − SMR Rating looks at a company's sales growth over the last three quarters, its before and after tax profit margins, and its return on equity. SMR Rating ranges from A − E, with A being the best and representing the top 20% of all companies. The B stocks are in the next 20% and so on. The E stocks represent the bottom 20% and the lesser quality companies.

Accumulation/Distribution (Acc/Dis) Rating The fifth proprietary rating, Accumulation/Distribution Rating, evaluates a stock's price and volume over the past 13 weeks to determine if a stock is going up in price on high volume (being bought by institutional investors), or going down in price on high volume (being sold by the institutions).

Where Is It Found? You'll find these ratings every day, exclusively in Investor's Business Daily's stock tables. A sample is presented in Figure 59.

A Word of Caution: Investor's Business Daily says, "SmartSelect" Corporate Ratings are the first step to identifying potential market leaders. IBD Stock Checkup (found on their Web site at www.investors.com) provides further evaluation of all the critical fundamental and technical data most affecting a stock's price. Daily Graphs and IBD Charts can assist in accurately timing your buys and sells."

Also See: Beta for a Security, Dividend Yield, Earnings Growth, Price–Earnings (P/E) Ratio.

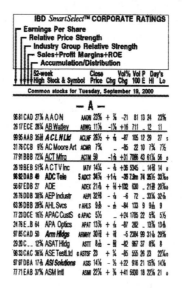

FIGURE 59 (From *Investor's Business Daily,* September 20, 2000, p. B8. With permission.)

TOOL #61

JAPANESE CANDLESTICK CHARTS

What Is This Tool? Candlestick charts use the same type of data as bar charts but construct each day's "candlestick" to emphasize the relationship between the day's open and close. They derive their name from the individual chart elements resembling a candlestick with wicks on other end. The thick bar of the candlestick is referred to as the real body and is determined by the day's open and close.

Traditionally in Japan, if the close is higher than the open, the body is colored red; however, this does not copy well, so it is common to leave it open–white. When the close is below the open, the body is filled—colored black. It is also common to refer to white (up) candlesticks as Yang and black (down) candlesticks as Yin.

The lines that extend from the top and bottom of the real body are called the shadow and represent the day's high and low trading range.

The close is also a period of great activity as day traders and "weak hands" unwind their positions. Weak hands refer to those traders who do not have great confidence in their positions and will easily reverse them. During periods of uncertainty, this will be done before the market closes, because these traders are afraid to suffer losses while the markets are closed when they cannot trade out of a position.

Patterns and signals from Japanese candlesticks tend to focus on determining trend reversal points without forecasting the magnitude or price objectives of these new trends. Some of these signals can be generated from a single day's candle, while most require two to three days to develop.

How Is It Computed? Each candlestick unit represents the trading activity for one time period. The body of the candlestick is defined by the opening and closing prices of the trading period. The close is above the open if it was higher than the open and the body of the candlestick is colored white (left empty). The close defines the bottom of the body if the closing price was below that of the open and the body is colored black (filled). The small vertical lines that extend above and below the body are called shadows. They represent the high and low trading range for the trading period.

Some patterns change in meaning and name depending upon whether they occur in an up or a down market. A hanging man in a market top closely resembles the market bottom signal of a hammer. Other chart patterns closely resemble Western patterns. The three Buddha's chart looks like the familiar head and shoulders pattern. (See Figure 60.)

Where Is It Found? Numerous computer software provide Japanese candlestick charts, such as MetaStock (800-882-3040) and Omega Research (800-328-1267).

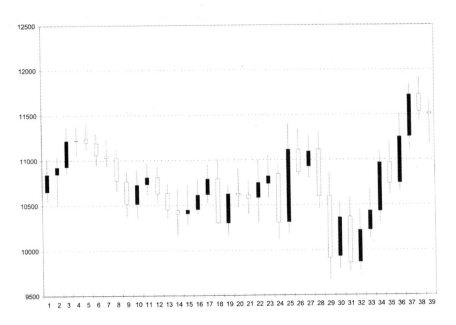

FIGURE 60   How the Dow Jones Industrial Average looks on a Japanese candlestick chart.

How Is It Used for Investment Decisions?   The futures market is well-suited to candlestick charting because the open and close are periods of great activity. Overnight news and rumors help to determine opening price, and often result in an opening price noticeably different from that of the close. In equity markets, the opening price tends to closely match that of the close unless a dramatic event has occurred while the markets were closed.

    A Word of Caution: Candlestick charts can, however, be used for equities. In fact, the Japanese use the chart to trade Japanese equities. Some individuals also recommend candlestick charts to help trade options. There are certain patterns that forecast increases and decreases in market volatility. Such changes in volatility would impact upon an option's premium and options traders could establish positions to benefit from such changes.

    Also See: Charting.

Tool #62

Japanese Yen

What Is This Tool?   As Japan's export business has soared, the yen has become one of the world's most important currencies. Its relationship to the U.S. dollar is a key to the globe marketplace and is seen as a barometer of Japan's economic strength vs. that of the United States.

How Is It Computed?   It is typically quoted in newspapers and financial reports on television and radio in terms of its relationship to the U.S. dollar. If the yen is at 135, that means that each U.S. dollar buys 135 yen. To figure out what 1 yen equals in U.S. currency, investors use this formula.

$$1 \text{ yen} = \$1 \text{U.S.}/\text{Yen-to-dollar rate}$$

Example   If $1 buys 135 yen, then 1 yen is equal to $1 divided by 135 or $.00741.

Where Is It Found?   Currency rates are listed daily in most major metropolitan newspapers as well as national publications like *The New York Times, USA Today,* and *The Wall Street Journal,* and on computer services such as America Online.

How Is It Used for Investment Decisions?   For American investors buying Japanese securities, the yen's movement is a key part of the profit potential in the investment. If the yen rises after an investment in Japanese securities is made, the value of those stocks or bonds to a U.S. investor will get a boost from the currency. That is because when the investment is sold, the stronger yen will generate more dollars when the proceeds are converted to U.S. currency. Conversely, a weak yen will be a negative to a Japanese investment for a U.S. investor.

In some cases, the movement of the yen also can also be viewed as an indicator of Japanese economic health. A strong yen can signal a buoyant economy, a possible indication to buy Japanese stocks. The yen's strength, however, should be verified not only against the U.S. dollar but against other major currencies.

A Word of Caution: The yen's strength vs. the dollar can be distorted by prevailing interest rates in each country. For example, a movement by Japan's central bank to stimulate the economy by cutting Japanese interest rates (two potential boosts for stock prices) could also cut the yen's price vs. the dollar if U.S. rates are stagnant or rising. This scenario of a falling yen might give an incorrect reading on the potential for buying Japanese securities.

Widely quoted currency rates are typically for transactions of $1 million or more. Consumers looking to use such figures to determine currency rates for foreign travel should expect to get somewhat less favorable exchange rates.

Also See: British Pound, German Deutsche Mark.

TOOL #63

JENSEN'S PERFORMANCE MEASURE (ALPHA)

What Is This Tool?   The Jensen measure called alpha, involves an ordinary least-squares regression to calculate risk-premium for an asset's return and the market return. A positive alpha means that the asset performed better than the market (e.g., Standard & Poor's 500) in risk-adjusted terms. A negative alpha means the opposite. If alpha is zero there is equality of return between the asset and the market on a risk-adjusted basis.

How Is It Computed?   Jensen's formula is $(r_i - r_f) = a + b(r_m - r_f)$ where

$(r_i - r_f)$ equals risk premium for portfolio i or asset i.
r_f equals risk-free rate.
r_m equals market return.
b equals beta coefficient.
a equals Jensen's performance measure (also termed *alpha*).
r_i equals return on portfolio i or asset i.

The alpha is computed with the ordinary least-squares regression subject to a sampling error. The alpha may or may not be statistically significant.

Jensen's alpha may be used to rank the performance of a portfolio when it is appropriately adjusted. The adjustment involves dividing each asset's alpha by the beta coefficient. For example, if assets X and Y are being ranked, use the ratio a/b.

Example: The ABC Fund, T-bills, and the Standard & Poor's 500 had the following returns over the previous five years.

	ABC Fund	T-Bills	S&P 500
Year	% Return	% Return	% Return
19X5	9	3	6
19X6	−4	8	−3
19X7	12	4	8
19X8	10	3	6
19X9	16	6	12

The alpha and beta coefficients for ABC Fund over the five-year period can be computed as follows. The following model of the ordinary least-squares regression is used.

$$(r_i - r_f) = a + b(r_m - r_f)$$

The variables for the regression in risk-premium terms are

Y variable equals $(r_i - r_f)$.
X variable equal $(r_m - r_f)$.

The following equation should be computed:

$$Y = a + bX$$
$$a = \overline{Y} - b\overline{X}$$

where

$$\overline{Y} = \Sigma Y/n$$
$$\overline{X} = \Sigma X/n$$

$$b = \frac{n\Sigma XY - (\Sigma X)(\Sigma Y)}{n\Sigma X^2 - (\Sigma X)^2}$$

Year	r_i ABC Fund % Return	r_f T-Bills % Return	r_m S&P 500 % Return	$(r_m - r_f)$ X	$(r_i - r_f)$ Y	XY	X^2
19X5	9	3	6	3	6	18	9
19X6	-4	8	-3	-11	-12	132	121
19X7	12	4	8	4	8	32	16
19X8	10	3	6	3	7	21	9
19X9	16	6	12	6	10	60	36
				5	19	263	191

Substitute these values into the formula for b first.

$$b = \frac{n\Sigma XY - (\Sigma X)(\Sigma Y)}{n\Sigma X^2 - \Sigma X^2} = \frac{(5)(263) - (5)(19)}{(5)(191) - (5)^2} = \frac{1220}{930} = 1.312$$

$$a = \overline{Y} - b\overline{X} = (19/5) - (1.311)(5/1) = 3.8 - (1.311)(1) = 2.488$$

Because of the positive alpha ($a = 2.49$), the ABC Fund performed better than the market in risk-adjusted terms for the time period involved.

Note: Using Excel regression yields the following

Summary Output

	Coefficients
Intercept	2.48817204 ← a
X	1.31182796 ← b

Example: An investor invests in the following stocks whose alpha and beta coefficients follow.

Stocks	a (alpha)	b (beta)
DEF Company	1.10%	0.80
LMN Company	1.30%	1.20
PQR Company	1.05%	0.90

The stocks would be ranked in performance by the Jensen measure when *a* (alpha) was divided by *b* (beta) as follows:

$$DEF = 1.10\%/0.8 = 1.375\% \quad \text{(second)}$$
$$LMN = 1.30\%/1.20 = 1.083\% \quad \text{(third)}$$
$$PQR = 1.05\%/0.90 = 1.676 \quad \text{(first)}$$

The positive values for Jensen's alpha reflect an underpricing of the security and, thus, an investment is financially attractive on a risk-adjusted basis.

Where Is It Found? The measure is found in Morningstar Inc. mutual fund publications. Investors may compute it themselves based on information extracted from financial publications including *Barron's* and *The Wall Street Journal.*

How Is It Used for Investment Decisions? The Jensen formula can be used to appraise the performance of a portfolio or asset.

Funds with positive and statistically significant alpha values can be said to outperform the market. On the other hand, funds with negative and statistically significant alpha values can be said to underperform the market. Funds with alphas not significantly different from zero perform equal to the market.

A Word of Caution: Measure is restricted to investors with some mathematical abilities.

Also See: Mutual Funds: Alpha for a Mutual Fund.

TOOL #64

LIPPER MUTUAL FUND INDEXES

What Are These Tools? These are daily benchmarks for stock mutual funds compiled by Lipper Analytical Services of New Jersey. These indexes show the value (or lack thereof) of various mutual fund portfolios.

How Are They Computed? Each member fund contributes an equal weighting to its respective index. As of November 1998, Lipper compiled 61 fund indexes in four broad categories.

- General equity, including growth and S&P 500 index funds (8 indexes).
- Specialized equity, including utility and international funds (12 indexes).
- Other equity, including convertible securities and balanced funds (5 indexes).
- Fixed income, including Treasury, corporate, and muni-bond funds (36 indexes).

Where Are They Found? Daily readings of these indexes are in some newspapers. A weekly summation is found in *Barron's.* (See Figure 61.)

How Are They Used for Investment Decisions? The indexes give investors a reading of how mutual fund managers are faring, especially when their results are compared to broad market indexes. Using the indexes, an investor also can gauge how his mutual fund, or his own portfolio, is faring against a "batting average" of top-paid professionals.

The index also can be used as a benchmark for growth-oriented stocks, issues of companies that display above average potential for rising sales and profits. When this index is outpacing a more cyclical index like the Dow Jones Industrial Average, it can be viewed as a signal that investor sentiment is turning toward growth issues.

A Word of Caution: Many mutual fund managers use various management styles even within a single portfolio or turn to cash in particularly rough markets. In some cases, the fact that a Lipper index outperforms the Dow Industrials may not be a sign of a certain style's strength. For example, it could be that skittish managers are holding large amounts of money out of a falling market.

Also See: Lipper Mutual Fund Rankings.

Lipper fund indexes

Type of Lipper index	Total return[1]		
	Thurs.	4 weeks	2000
Balanced	0.7%	3.4%	6.0%
Emerging market	0.1%	1.7%	−12.4%
Equity income	0.7%	3.6%	4.1%
Gold	1.2%	6.0%	−14.7%
International	0.1%	3.4%	−5.7%
Science & technology	2.4%	16.0%	18.8%
Large-cap core	1.1%	5.6%	7.2%
Large-cap growth	1.5%	8.0%	6.7%
Large-cap value	0.8%	3.8%	3.2%
Midcap growth	2.6%	16.1%	13.1%
Midcap value	0.6%	5.2%	9.3%
Multi-cap growth	1.7%	9.9%	15.3%
Multi-cap value	0.7%	4.4%	6.4%
Small-cap growth	1.5%	11.3%	15.6%
Small-cap value	0.6%	5.6%	12.5%

1 – Capital gains and dividends reinvested   Source: Lipper

FIGURE 61   How a newspaper tracks Lipper mutual fund indexes. (Copyright 2000, USA TODAY. Reprinted with permission.)

Tool #65

Lipper Mutual Fund Rankings

What Are These Tools?   These are widely watched measures of how mutual funds perform against each other, both on an industrywide and peer-group basis, as compiled by Lipper Analytical Services of New Jersey. Lipper issues reports weekly, monthly, and quarterly. Its long-term rankings with track records of 1 to 15 years are the most quoted.

How Are They Computed?   Lipper tracks more than 8000 funds: stock funds, bond funds, and money markets. The Lipper rankings are based strictly on total return, which is price appreciation or depreciation plus dividends paid out. No further analysis is done. That is unlike rankings by Morningstar Inc. that consider funds for both return, consistency of that return, and risk taking. Lipper then splits funds into various categories that range from broad ones (stocks, bonds, money funds) to extremely narrow ones (international stock funds by specific country).

Where Are They Found?   Lipper results are the basis for many newspapers' weekly and quarterly coverage of the fund industry. Every business day, *The Wall Street Journal* uses Lipper data to highlight the best and worst funds in a specific fund category. Lipper's report on the quarterly performance of stock funds is issued within a week of a quarter's end and gets nationwide media coverage. The Lipper performance numbers also frequently appear in mutual fund advertising. (See Figure 62.)

How Are They Used for Investment Decisions?   The Lipper numbers often are used to rate one mutual fund against another. Investors can compare their funds to its peers for total return performance.

There are also the often discussed Lipper industrywide rankings which, for example, propelled Peter Lynch of Fidelity Investments into prominence. During his 15-year tenure on the Fidelity Magellan Fund, no fund manager beat Lynch's total return figures as compiled by Lipper.

Lipper's fund category rankings can help investors who want to invest in narrow sector funds that limit assets to one industry or country. Groups with rising performance figures could be an investment while a category taking on losses could be ripe for selling.

A Word of Caution: Experts suggest that investors look at long-term track records of 5 years or more when choosing among funds. They also advise that past performance figures like Lipper's are no indication of what the future may bring.

Also See: Morningstar Mutual Fund Rankings.

■ BEST OF THE BUNCH

This exclusive feature of The Orange County Register allows you to survey 45 groups of mutual funds and track the two top-performing funds during three time periods ended last week, based on data from Lipper Analytical Services. Performance is based on total return — price change plus dividend. Footnotes: **n:** no-load fund; **r:** redemption fee; **t:** distribution fee; **p:** redemption and distribution fees; **x:** ex-dividend. **FREE INFORMATION:** Free financial information and prospectuses available wherever the ♣ symbol appears by calling (888) 219-7044 or faxing (888) 202-3209, or to download immediately, visit **ocregister.fundclub.com**. Information is applicable to U.S. addresses only and will be sent the next business day via first-class mail, subject to availability. Sponsored by participating funds.

U.S. stock funds

Objective	Best fund last 3 months	Pct. gain	Best fund last 12 months	Pct. gain	Best fund last 3 years	Pct. gain
Gold (AU): Precious metal investments	Invesco: Gold np	+4.3%	First Eagle SoGen: Gold	+7.6%	Franklin Class A: GoldPrM A	-29.0%
	Scudder Funds: ♣Gold n	+2.7%	Fidelity Selects: Gold r	+5.6%	Oppenheimer A: Gold p	-30.7%
Equity Income (EI): Income-producing stocks	SG Cowen Funds: IncGrA px	+22.5%	AmSouth Fds Trust: EquityInc	+25.5%	Parnassus Funds: EqtyInco n	+69.0%
	MFS Funds A: EqtyIncA p	+17.7%	Rockhaven Fds: Fund p	+25.4%	AmSouth Fds A: EqIncome	+61.6%
Health/Biotech (HB): Medical stocks	Fidelity Selects: MedDel r	+24.6%	Dresdner RCM: BiotechN p	+230.3%	Fidelity Selects: Biotech r	+227.1%
	Vanguard Fds: HlthCare n	+17.5%	Franklin Class A: BioDisA p	+192.1%	Dresdner RCM: GlobHlth p	+164.4%
Large-Cap (LC): Varied, big U.S. companies	Van Kamp Funds A: ExchFd	+13.4%	Whitehall Funds: ♣Growth	+56.6%	Whitehall Funds: ♣Growth	+146.1%
	SEI Portfolios: CapApA n	+12.5%	SunAmerica Focus: FocusA p	+45.8%	BNY Hamilton Instit: LgCapGth n	+107.0%
Large-Cap Growth (LB): Big, growing U.S. firms	Oak Assoc Fds: WhiteOakG n	+13.2%	Berkshire Funds: Focus	+168.3%	PBHG Funds: LrgCap20	+333.5%
	WPStewGr	+10.6%	PBHG Funds: LrgCap20	+113.5%	Rydex Investor: OTC n	+285.2%
Large-Cap Value (LV): Big, out-of-favor U.S. firms	Fidelity Advisor I: DivGrthl	+15.3%	ARK Funds: BlueChipEqI	+23.5%	ARK Funds: BlueChipEqI	+86.1%
	Fidelity Advisor A: DivGrowthA	+15.2%	ARK Funds: BlueChEqA p	+23.3%	ARK Funds: BlueChEqA p	+85.2%

FIGURE 62 How a newspaper uses Lipper mutual fund rankings. (From *Orange County Register* (CA), June 18, 2000, p. 5. With permission.)

Tool #66

Misery Index

What Is This Tool?   This is an index that tracks economic conditions including inflation and unemployment. It was particularly referred to in the economically depressed period of 1977 through 1981 in the United States. The inflation rate was in the double digits at that time.

How Is It Computed?   The Misery Index equals inflation rate plus unemployment rate plus the prime rate.

Where Is It Found?   Information needed to compile this index is found in most newspapers and business magazines but may be easiest to locate in computer databases like America Online, where old news stories are often simple to find.

How Is It Used for Investment Decisions?   The index typically is negatively correlated to the current condition of the stock market.

A Word of Caution: The misery index is not useful in predicting future stock prices.

Tool #67

Momentum Gauges

OSCILLATORS

What Are These Tools?   Oscillators are an internal measure (strength) of momentum. They are used to measure a price index by looking at the rate of change (ROC). They are a short- to intermediate-term breadth indicator. This technical analysis method is used to obtain historical comparative information. Oscillators reflect movements of similar proportion in the same way. An example of a momentum oscillator is the advance–decline line.

How Are They Computed?   Momentum is measured by computing the rate of change in a market average or index over a prescribed time period. It is calculated by subtracting a 39-day exponential moving average of the net difference between the number of advancing issues and the number of declining issues from a 19-day exponential moving average of the net difference between the number of advancing issues and the number of declining issues. New York Stock Exchange data is used in the calculation.

Example: To construct an index measuring a 26-week rate of change, the current price is divided by the price 25 weeks ago. If the current price is 150 and the price 25 weeks ago was 160, the rate of change (momentum index) is 93.8 (150/160). The next reading in the index would be determined by dividing next week's price by the price 24 weeks ago.

Where Are They Found?   Oscillators are found in brokerage research reports prepared by technical analysts.

How Are They Used for Investment Decisions?   Oscillators can show the rhythm in price movement, which could aid the investor in determining the degree of momentum associated with stocks. Are stocks in a down or an up cycle? Strength in stock price is a bullish indication while weakness in stock prices is a bearish sign.

The oscillator signals when the market is overbought or oversold. The oscillator typically reaches an extreme reading prior to a change in the trend of stock prices.

The bear market selling climax is indicated by an oscillator reading of approximately −150. A surge of buying activity, implying an overbought condition, is signaled by an oscillator reading above 100.

The oscillator normally passes through zero near market tops and bottoms. When the oscillator goes from below to above zero, it is considered bullish for stock prices. On the other hand, when it goes from above to below zero, it is interpreted as bearish for the stock market.

Also See: Momentum Gauges: Summation Index.

RELATIVE STRENGTH ANALYSIS

What Is This Tool?   Relative strength is the price performance of a company's stock compared to the price performance of an overall market and/or industry index, In addition, the investor can compare how the performance of the stocks within an industry has done relative to the overall market. Market indexes that might be used are the Dow Jones Industrial Average, Standard & Poor's 500, and New York Stock Exchange Composite. Relative strength is an approach used in technical investment analysis by chartists.

How Is It Computed?   Relative strength for a company's stock may be computed using one or both of the following ratios:

$$\frac{\text{Monthly average stock price}}{\text{Monthly average market index}}$$

or

$$\frac{\text{Monthly average stock price}}{\text{Monthly average industry group index}}$$

An increase in these ratios means that the company's stock is performing better than the overall market or industry. This is a positive sign. A relative strength index (RSI) also may be computed for a security as follows.

$$\text{RSI} = 100 - \frac{100}{1 + \text{RS}}$$

where

$$\text{RS} = \frac{\text{14-day average of up closing prices}}{\text{14-day average of down closing prices}}$$

Relative strength for an industry also may be determined as

$$\frac{\text{Specific industry group price index}}{\text{Total market index}}$$

An increasing ratio indicates that the industry is outperforming the market.

Where Is It Found?   Brokerage research reports, *Value Line Investment Survey,* and new stock reports from Morningstar Inc. may provide relative strength information on companies and industries. Investors also may compute the monthly average price of a company's stock and that of a market index by referring to price quotations in newspapers.

How Is It Used and Applied?   Relative strength is an approach that helps the investor determine the quality of a price trend of an index or stock by comparing it with the trend in another index or stock. If there is an improvement in relative strength after a drastic decline, this is an indicator of strength. If there is a deterioration in relative strength after a prolonged increase in price, weakness exists.

An industry index (e.g., utilities) may historically lead an overall market index (e.g., Dow Jones Industrial Average) so that a change in that industry index may infer a future change in the overall market index. A graph may be prepared charting these indexes relative to one another. (See Figure 63.)

How Is It Used for Investment Decisions?   The investor can evaluate relative strength to predict individual stock prices. If a stock or industry group outperforms the market as a whole, the investor should view that stock or industry positively. The presumption is that strong stocks or groups will become even stronger.

A high relative strength index for a security (stock or bond) or commodity indicates an overbought situation while a low RSI indicates an oversold environment.

A Word of Caution: A distinction must be made between relative strength in a contracting market and relative strength in an expanding market. When a stock outperforms a major stock average in an advance, it possibly may soon turn around, but when the stock outperforms the rest of the market in a decline, the stock will likely remain strong.

Relative Strength for Stocks

Company Name	Ticker	Exchange	1-Year Total Return	1-Year Relative Strength
Yahoo!	YHOO	NASDAQ	497%	397
Amazon.com	AMZN	NASDAQ	315%	245
America Online	AOL	NYSE	231%	175
Dell Computer	DELL	NASDAQ	227%	172
Lycos	LCOS	NASDAQ	211%	159
Excite	XCIT	NASDAQ	209%	157
Infoseek	SEEK	NASDAQ	179%	132
EMC	EMC	NYSE	130%	91
Apple Computer	AAPL	NASDAQ	118%	81
Wal-Mart Stores	WMT	NYSE	98%	64
Lucent Technologies	LU	NYSE	95%	62
Gateway 2000	GTW	NYSE	92%	60
Ford Motor	F	NYSE	95%	56
Staples	SPLS	NASDAQ	86%	55
Tele-Communications A	TCOMA	NASDAQ	84%	53
Comcast	CMCSK	NASDAQ	80%	49
Ascend Communications	ASND	NASDAQ	79%	49
Novell	NOVL	NASDAQ	76%	47
Cisco Systems	CSCO	NASDAQ	73%	44
Sun Microsystems	SUNW	NASDAQ	70%	42

FIGURE 63   A look at the highest 1-year relative strength figures and the total return (price appreciation plus dividends, if any) for actively traded stocks (average daily volume of 2 million or more) for the year ended October 31, 1998.

SUMMATION INDEX

What Is This Tool?   The McClellan Summation Index is simply a cumulative total of the McClellan Oscillator. It is used for interpreting intermediate- to long-term moves in the stock market.

How Is It Computed?   This is simply a cumulative total of the McClellan Oscillator.

Where Is It Found?   It is frequently reported on the cable TV network *CNBC*.

How Is It Used for Investment Decisions?   Similar to the McClellan Oscillator, the McClellan Summation Index gives buy and sell signals. When it crosses zero, it is considered bullish for stock prices. When it crosses from above to below zero, it is interpreted as a bearish sign for the stock market.

A Word of Caution: This index reading is far from an exact science.

Also See: Momentum Gauges: Oscillator.

"WALL STREET WEEK" TECHNICAL MARKET INDEX

What Is This Tool?   This index is based on a survey of 10 technical investment analysis methods. These technical indicators include financial conditions, market activity, investor behavior, insider purchases, calls and puts, and monetary policy. The index is used to substantiate an upward or downward trend in the stock market. Fundamental analysis is not taken into account.

How Is It Computed?   This measure is based on assigning +1 for a bullish characteristic, −1 for a bearish situation, and 0 for no effect for each of the 10 technical indicators. The ratings are then added to obtain a total. The 10 indicators in the index are

1. *Market Breadth*—a moving average for 10 days for the net effect of advancing issues relative to declining issues.
2. *Put/Call Options Premium*—a ratio of premium on put options to premium on call options.
3. *Arms Short-Term Trading Index*—an advance–decline ratio of "Big Board" stocks shown by the rate of volume increasing to volume decreasing.
4. *Insider Buy–Sell Ratio*—a ratio of insider buys to insider sells.
5. *Low-Price Activity Percentage*—the volume of low-priced "risky" securities to the volume of high-quality stocks.
6. *Bearish Sentiment Index*—an indicator of how investment newsletters perceive the future condition of the stock market. It is determined by Investor's Intelligence.
7. *DJIA Momentum Ratio*—the difference between the closing DJIA and the average DJIA for 30 days.
8. *NYSE High–Low Index*—the number of stocks accomplishing new highs relative to those reaching new lows over the previous 10 trading days applied on a daily basis.

9. *NYSE Securities at Market Prices above 10-Week and 30-Week Moving Averages*—the percentage of stocks selling above their 10-week and 30-week highs.

10. *Ratio of Ending Prices on Fed Funds to the Discount Rate*—when the Federal Reserve Board tightens the money supply, the rate of Fed increases relative to the discount rate. This is because the Federal Reserve Board is charging a higher rate between member banks and a lower rate (discount rate) is being charged by the Federal Reserve for member banks to borrow from the Federal Reserve.

The upper and lower limits for the Wall Street Week Technical Market Index is + 10 to −10. A reading of + 1 to −1 is neutral. A reading of +5 or more is a buy signal while a reading of −5 is a sell signal.

Where Is It Found?   The index is found in *Futures* and *Investor's Analysis* [505-820-2737] (published by Robert Nurock) magazine.

How Is It Used for Investment Decisions?   The index indicates, from a technical standpoint, whether the bottom or top is indicated. At market bottom, there is a buy indicator while at market top there is a sell indicator. For example, at the market bottom stock prices are expected to increase.

Tool #68

Morgan Stanley EAFE Index

What Is This Tool?   The Morgan Stanley Europe, Asia, and Far East Index is considered the key "rest-of-the-world" index for U.S. investors, much as the Dow Jones Industrial Average is for the American market. The index is used as a guide to see how U.S. shares are faring against other markets around the globe. It also serves as a performance benchmark for international mutual funds that hold non-U.S. assets.

Morgan Stanley also compiles indexes for most of the world's major stock markets as well as for many smaller, so-called "emerging" markets. In addition, there are Morgan Stanley indexes for each continent and the entire globe.

How Is It Computed?   First, Morgan Stanley has created its own indexes for 18 major foreign markets. To make the EAFE Index, those country indexes are weighted to reflect the total market capitalization of each country's markets as a share in the world market.

The index is quoted two ways: one in local currencies and a second in the U.S. dollar. This shows how American investors would fare addressing both share price and currency fluctuations.

Where Is It Found?   The EAFE Index can be found in newspapers such as *Barron's.*

How Is It Used for Investment Decisions?   The EAFE index can be used by investors to gauge the proper exposure to foreign investments. Historically, foreign shares have produced slightly better results than U.S. issues, particularly in the past two decades because the American economy has matured and overseas industries are now the fastest growing.

When the EAFE Index is performing better than the U.S. markets, it may be time for investors to shift money overseas. Conversely, when U.S. market indexes are doing better than the EAFE Index, a shift away from foreign assets may be in order.

A Word of Caution: Currency fluctuations can play a major part of any overseas investment. A rising EAFE may be more a reflection of a weak U.S. dollar than improving foreign economies or strong opportunities in overseas stocks. (See Figure 64.)

Also See: FTSE 100, Frankfurt Dax Index, Nikkei Stock Index.

Other MSCI Foreign Stock Indexes

- *MSCI World Index*—market value-weighted equity index of over 1500 company issues traded in 22 world markets.
- *MSCI Emerging Markets Free Index*—market value-weighted barometer of over 850 company issues traded in 22 world markets.
- *MSCI Emerging Markets Free/Latin America Index*—market value-weighted barometer of approximately 170 company issues traded in 7 Latin American markets.
- *MSCI Europe Index*—market value-weighted barometer of over 550 company issues traded in 14 European markets.
- *MSCI Far East excluding Japan Free Index*—market value-weighted barometer of over 450 company issues traded in 8 Asian markets, excluding Japan.
- *MSCI France Index*—unmanaged index of over 75 foreign stock prices, converted into U.S. dollars, assuming reinvestment of all dividends paid.
- *MSCI Germany Index*—unmanaged index of over 75 foreign stock prices, converted into U.S. dollars, assuming reinvestment of all dividends.
- *MSCI Hong Kong Index*—unmanaged index of over 38 foreign stock prices, converted into U.S. dollars, assuming reinvestment of all dividends paid.
- *MSCI Japan Index*—unmanaged index of over 317 foreign stock prices, converted into U.S. dollars, assuming reinvestment of all dividends paid.
- *MSCI Nordic Countries Free Index*—unmanaged index of over 95 foreign stock prices, converted into U.S. dollars, assuming reinvestment of all dividends.
- *MSCI Pacific Index*—market value weighted barometer of over 400 company issues traded in 6 Pacific-region markets.
- *MSCI United Kingdom Index*—unmanaged index of over 143 foreign stock prices, converted into U.S. dollars, assuming reinvestment of all dividends paid.

FIGURE 64 Morgan Stanley Capital International, which tracks the widely quoted EAFE Index tracking non-American shares, has other global stock indexes as well.

Tool #69

Morningstar Mutual Fund Rankings

What Are These Tools?   A risk measurement system for comparing more than 2000 mutual funds' long-term performance is available from Chicago-based Morningstar Inc. The system rates stock and bond funds from five stars (the best) to one star (the worst). In some cases, funds are unrated.

How Are They Computed?   Morningstar uses a proprietary system that measures a mutual fund's price and dividend performance as well as the risks taken by the fund management to get those results.

The rankings are then made from comparing a fund both in its own category and against the industry as a whole. Thus, the best performing fund in a category that has been in weak market sector might get only two or three stars.

As of 1999, Morningstar assigns the top 10 percent of an asset class (international equity, domestic equity, taxable bond, or municipal bond) with 5 stars (highest). Those falling in the next 22.5 percent receive 4 stars (above average); a place in the middle 35 percent earns 3 stars (neutral); those in the next 22.5 percent, receive 2 stars (below average); and the bottom 10 percent receive get 1 star (lowest).

Morningstar, also, now offers so-called "style reviews" of funds, putting them in one of nine categories designated by what they own, not what their marketing department says the fund is.

The nine categories, called "style boxes," for stock funds combine both a fund's investment methodology and the size of the companies in which it invests. The nine categories are small cap, medium cap, or large cap each divided into growth, value, or blend slices.

The equity style box measures the size of the companies held in a fund's portfolio by median market capitalization and classifies funds as small cap, medium cap, or large cap.

A small-cap fund's median market capitalization must be less than $1 billion; median market caps, $1 billion to $5 billion; median market caps greater than $5 billion are classified as large cap.

To track "investment methodologies," Morningstar divides the average price-to-earnings ratio and price-to-book ratios of a fund by those of the S&P 500 and adds the results. Funds with a total less than 1.75 are classified as value-oriented; blend-oriented funds have a total between 1.75 and 2.25; and growth-oriented funds are greater than 2.25. (The S&P 500's sum is 2.00.)

For bond funds, the nine style categories are sliced by maturity (short-term, intermediate-term, and long-term) and credit-quality (high, medium, and low) to provide a snapshot of risk. So, a fixed-income fund in the short maturity/high quality style should be among the safest while funds in the long maturity/low quality style would be the riskiest. (See Figure 65.)

Where Are They Found?   Morningstar sells it rating service, Morningstar Mutual Funds, to the public in a print or computerized compact disc version. Some of the analysis is also contained on its web site in the Internet at www.morningstar.com. The printed version is sold in a loose-leaf book style much like Value Line's well-known rankings of individual stocks. The service updates each of its 10 parts every 6 months by issuing new reports on a rotating basis approximately twice a month. Morningstar rankings are also often discussed in the newspaper and magazine accounts of the funds industry. They are the basis for *The Los Angeles Times'* quarterly fund report.

How Are They Used for Investment Decisions?   When choosing among mutual funds, investors can use Morningstar rankings to find potentially better-performing investments. Many brokerages and financial planning firms limit their clients' investments to 5-star and 4-star funds.

However, choosing a 5-star fund over a 3-star fund is not always the correct choice. For one, Morningstar's rankings reflect past performance and that often slants the reviews toward funds with recently successful investment styles.

In addition, within each category (notably a poorly performing sector) the highest rated fund may have succeeded by limiting its exposure to certain risks. If an investor believed that an out-of-favor market sector was ready to return, he might want to buy a fund with a lower rating that was more fully invested in that sector.

A Word of Caution: Before buying or selling a fund, investors should consult other fund-watching sources. *Business Week, Forbes,* and *Money* magazines all print periodic analyses of individual funds as do many newsletters and newspapers. (See Figure 66.)

Also See: Lipper Mutual Fund Rankings.

FIGURE 65 Morningstar Style Boxes. Here is how Morningstar's "style reviews" are put in an easy-to-track grid.

Advanced Analytics	Page 1 of 1									Release Date: 09-30-1998		

Fidelity Magellan
Release Date: 09-30-1998

	Rating	Net assets	Morningstar Category
	★★★★	$65882.0 mil	Large Blend

Trailing-Time-Period Performance

	YTD	1 Mo	3 Mo	12 Mo	3 Yr Annlzd	5 Yr Annlzd	10 Yr Annlzd	15 Yr Annlzd	Load-Adjusted Return % as of 09-30-98	
Total Return % as of 09-30-98	5.03	6.05	-11.05	4.61	13.53	14.74	17.35	17.07	12 Mo	1.47
+/- S&P 500	-0.98	-0.36	-1.12	-4.45	-9.07	-5.16	0.07	0.65	5 Yr	14.04
+/- Wil Large Blend	-0.53	-0.63	-0.76	-4.15	-8.18	-3.78	0.65	1.29	10 Yr	17.00
% Rank within Category				46	89	70	9	2	Inception	—

1 = Best 100 = Worst

Calendar-Year Performance

	1988	1989	1990	1991	1992	1993	1994	1995	1996	1997	9-98
Total Return %	22.77	34.58	-4.51	41.03	7.02	24.66	-1.81	36.82	11.69	26.59	5.03
+/- S&P 500	6.16	2.90	-1.39	10.55	-0.60	14.60	-3.13	-0.71	-11.26	-6.76	-0.98
+/- Wil Large Blend	5.57	3.16	-0.43	8.56	-1.18	14.92	0.54	-0.84	-9.86	-6.71	-0.53

Operations

Family:	Fidelity Group	Ticker:	FMAGX	Front-End Load:	3.00%
Inception:	05-1963	Min Init Purchase:	Closed	Deferred Load:	0.00%
Manager:	Stansky, Robert E.	Min IRA:	Closed	12b-1 Fee:	0.00%
Tenure:	2 Years	Min Auto Inv Plan:	Closed	Expense Ratio:	0.61%
Telephone:	800-544-8888	Purchase Constraints:	C/	NAV:	97.52
Objective	Growth				

FIGURE 66A Here's how parts of Morningstar reports look.

PIMCo Total Return Instl	Rating	Net Assets	Morningstar Category
Release Date:09-30-1998	★★★★★	$18499.9 mil	Intermediate-term Bond

Investment Approach

Equity Style	Equity Portfolio Statistics	Portfolio Avg	Relative Index	Relative Category	Composition % of assets as of 06-30-98		Regional Exposure % of non-cash assets	
Value Blend Growth	Price/Earnings Ratio	—	—	—	Cash	16.4	U.S.&Canada	—
Large	Price/Cash Flow	—	—	—	U.S. Stocks	0.0	Europe	—
	Price/Book Ratio	—	—	—	Non-U.S. Stocks	0.0	Japan	—
Medium	3 Yr Earnings Growth %	—	—	—	Bonds	82.8	Latin America	—
Small	Med Market Cap ($mil)	—	—	—	Other	0.7	Pacific Rim	—
							Other	—

Fixed-Income Style	Fixed-Income Portfolio Statistics		Credit Analysis as of 06-30-98	
Short Int Long	Avg Eff Mat/Duration	11.3 Yrs/5.0 Yrs		% of Bonds
High	Avg Weighted Coupon	6.80		
	Avg Weighted Price	101.00	U.S. Government/Agency	39.00
Medium	Avg Credit Quality	AA	AAA	20.00
			AA	4.00
Low			A	21.00
			BBB	8.00

Turnover Rate	173%		BB	8.00
Assets in Top 10 Holdings %	41.38		B	0.00
Total Number of Holdings	2612		Below B	0.00
			Not Rated/Not Available	0.00

FIGURE 66B   Here's how parts of Morningstar reports look.

Tool #70

Mortgage Rates

WEEKLY AVERAGE

What Are These Tools?   These are a national average of interest rates offered on mortgages—both 30-year fixed and starting rates for 30-year adjustable loans that vary once a year—from surveys by the Federal National Mortgage Association, also known as Fannie Mae, and the Federal Home Mortgage Corporation, or Freddie Mac.

How Are They Computed?   Fannie Mae and Freddie Mac poll major lenders nationwide and compile an average for these two key mortgage products.

Where Are They Found?   The two agencies release the information every Thursday and Friday, and short stories about it appear in major daily newspapers. It is also found in the statistical tables of *Barron's*.

How Are They Used for Investment Decisions?   For those shopping for a mortgage, the averages can be reviewed to see how available loans compare. Movements in the weekly average can help loan shoppers decide when to lock in a rate on a new mortgage. (See Figure 67.)

In the big picture, mortgage rates are often seen as a good indicator of the housing industry. Falling rates can be a boost to home building, a key element of any economic recovery. However, low mortgage rates do not guarantee a pickup in home sales or building activity. One group that clearly will benefit from falling rates is lenders, who also can profit from mortgage refinancings.

Also See: Interest Rates: 30-Year Treasury Bonds.

ADJUSTABLE LOAN BENCHMARKS

What Are These Tools?   Various indexes serve as the base rate for variable mortgages. Indexes are selected by the borrower when the loan is made. Typically, lenders determine monthly payments by charging between 2 percentage points and 4 percentage points above the indexes.

How Are They Computed?   The indexes come from a variety of sources. One widely used is an odd average known as the 11th District Cost of Funds Index. This monthly index, compiled by the Federal Home Loan Bank of San Francisco, tracks monthly costs of deposits at savings and loans in California, Nevada, and Arizona.

Other popular benchmarks are figures compiled by the Federal Reserve known as constant maturity averages for various U.S. Treasury issue maturities. They are

▶ 30-year adjustable (conforming)

Institution	Phone	Start rate%	Down pmt%	Points %	APR %	Max. Samt	Lock days	Index	Margin	Init Adj	Reg. Adj	CAPS
Bank of America	800-843-2632	5.250	20	0.000	7.668	252,700	60	1TA	2.000	3M	1M	0.00/6.63
First Federal CA	800-672-4332	3.950	20	1.000	7.779	252,700	60	11D	2.600	1M	1M	0.00/8.00
Cal Fed Lending	800-225-3337	5.250	20	0.000	7.790	252,700	30	1TA	2.125	3M	1M	0.00/6.63
World Savings	800-333-4193	4.750	20	1.000	7.857	252,700	45	11D	2.650	1M	1M	0.00/7.20
PFF Bank & Trust	888-733-5465	4.000	20	1.000	7.983	252,700	45	1TA	2.250	3M	1M	0.00/6.00
Countrywide	800-877-5626	3.375	20	1.000	8.013	252,700	45	11D	2.875	3M	1M	0.00/9.38

▶ 30-year adjustable (jumbo)

Institution	Phone	Start rate%	Down pmt%	Points %	APR %	Max. Samt	Lock days	Index	Margin	Init Adj	Reg. Adj	CAPS
Bank of America	800-843-2632	5.250	20	0.000	7.678	750,000	60	1TA	2.000	3M	1M	0.00/6.63
First Federal CA	800-672-4332	3.950	20	1.000	7.780	1,000,000	60	11D	2.600	1M	1M	0.00/8.00
Cal Fed Lending	800-225-3337	5.250	30	0.000	7.801	1,000,000	30	1TA	2.125	3M	1M	0.00/6.63
World Savings	800-333-4193	4.750	20	1.000	7.857	600,000	45	11D	2.650	1M	1M	0.00/7.20
PFF Bank & Trust	888-733-5465	4.000	20	1.000	8.005	650,000	45	1TA	2.250	3M	1M	0.00/6.00
Downey Savings	800-348-5931	3.850	20	0.250	8.049	600,000	30	11D	2.950	1M	1M	0.00/6.10

FIGURE 67   How a newspaper tracks mortgage rates. (From *Orange County Register* (CA), June 18, 2000, p. 5. With permission.)

reported weekly. Another benchmark is the London Interbank Offered Rate or LIBOR. It is reported daily and reflects what the world's bankers charge one another for money.

Where Are They Found?   Many newspapers' business or real estate sections such as *The Dallas Morning News, The Los Angeles Times, The New York Times,* and *Orange County Register,* now track the more widely used benchmarks weekly. Major adjustable loan lenders can also be contacted directly for periodic updates on the indexes. (See Figure 68.)

How Are They Used for Investment Decisions?   A borrower considering an adjustable loan must juggle several factors, notably which index to choose.

History shows that no one index stands out as the best for the consumer. The 11th District index is known for its slow change while Treasury or LIBOR benchmarks can be more volatile. Movements of these benchmarks can show how fat—or stretched—the homeowner's wallet is. That is not a bad economic indicator.

Also, mutual funds that concentrate on buying securities backed by adjustable rate mortgages are lower-risk alternatives to money market mutual funds and savings accounts. Trends in these benchmarks would show up in the yields of such funds just a few months in the future.

► Indexes for adjustables			
	This week	Last week	Year Ago
11th Dist Cost of Funds (11D)	(4/00)-5.078%	(3/00)-5.002%	(4/99)-4.490%
6-month Trea.(6TB)	6.01%	6.04%	4.81%
1-year Treas. (1TS) (wkly avg. yield-CMT)	6.23%	6.30%	5.12%
1-year Treas. (1TA) (12 mo. moving avg.)	5.703%	5.703%	4.784%
Six-month LIBOR (6LB)	6.960%	6.980%	5.375%
One-month LIBOR (1LB)	6.650%	6.630%	5.000%
Federal Nat'l Mtg. Assoc. (par, 30-yr fixed,30day)	8.25%	8.32%	7.68%
Prime rate (PRI)	9.50%	9.50%	7.75%
6-month CD (6CD)	6.92%	7.02%	5.24%
Sources: Federal Reserve Statistical Release H-15; FHLBB; Mort. News Co.			

FIGURE 68 Here's a look at how mortgage benchmarks appear in a newspaper. (From *Orange County Register* (CA), June 18, 2000, p. 5. With permission.)

Tool #71

Most Active Issues

What Is This Tool? The most active issues are the stocks that have the largest share volume for the trading period, usually daily. They are listed in the order of shares traded. Active stocks typically reflect the activity of institutional investors, but a diminutive-yet-popular issue—like Internet-related issues in 1998—can crack the list based on heavy small investor activity.

How Is It Determined? A tabulation is made of the stocks of each market in terms of trading volume based on the number of shares that changed ownership.

Where Is It Found? Statistics on the most active stocks are published in the financial press on both a daily and weekly basis. For example, *The New York Times* lists the most active issues for the trading day on the NYSE, AMEX, and NASDAQ markets. It lists the volume, last price, and change. *Barron's* lists the high–low–last prices and change. America Online and other online computer database services list the most active issues on the major exchanges as well. Each market's Web site (www.nyse.com; www.nasdaq.com; www.amex.com) also tracks this data. (See Figure 69.)

How Is It Used and Applied? The quality of the volume leaders will be a key indicator. If they are blue chip companies, and if the active shares are up in price, the future might be bright. If they are secondary issues, or these market leaders are falling in price, investors should be somewhat cautious.

How Is It Used for Investment Decisions? The most active issues are those involving the greatest share volume because of their interest to institutional buyers

MOST ACTIVE ISSUES

NYSE Stock	Vol.	Cls.	Chg.	NASDAQ Stock	Vol.	Cls.	Chg.	AMEX Stock	Vol.	Cls.	Chg.
AT&T	25052400	32.00	+1.13	ADC Tel s	37551500	39.69	+3.63	DevonE	12485900	60.50	+.06
Lucent	24675700	44.38	+1.94	Microsft	28886200	70.94	-.38	Nasd100T s	11161500	99.13	+.69
Motorola s	18069200	36.00	+.50	DellCptr	28599290	40.56	+1.06	SPDR	3561900	151.80	+.03
Compaq	11392400	34.06	-.31	Cisco s	25133700	66.56	+.50	IvaxCp s	2789600	31.69	+1.31
FordM n	11211600	25.56	-.94	WrldCom s	22554500	36.63	-.56	Avanir n	1187300	4.50	-.13
Pfizer	9698400	42.34	-.53	Intel s	21821500	74.06	+.19	SP Mid	887000	98.44	+.44
SFeSnyder	9394000	13.25	-.06	Oracle	18422500	87.75	+1.00	Nabors	791700	47.27	-.35
NortelNw	8685100	80.00	-1.63	NextelC s	15091900	48.38	-3.00	BiotechT n	663700	192.50	-1.50
GenElec s	8178900	59.81	-.13	Yahoo s	13407300	121,00	-1.06	Xcelera s	650400	16.25	-.25
Xerox	7823500	17.00	-.13	GlblCrss	12304100	29.55	-2.08	SemiHTr n	608600	97.75	-.75

FIGURE 69 How a newspaper covers most actively traded stocks. (From *Miami Herald*, August 30, 2000, p. 5C. With permission.)

and sellers of stock. Such issues are highly marketable. In addition to looking at share volume, the investor must consider price movement. If the number of shares traded on a company's stock is very high but its market price is constant, the security is fairly stable in price. However, increasing price on heavy volume is significant. The security may be making a significant upward movement. On the other hand, decreasing price on heavy volume might be alarming as it may signal a significant downward movement—the "Big Boys" are unloading. The investor must look out for substantial price changes coupled with heavy trading activity.

The investor should determine the reason and significance of the heavy trading. Is it owing to an announcement of superior earnings, heavy buying by institutions, or a takeover attempt? Can the volume and price movement be sustained? Some company stocks such as General Electric, Microsoft, and Philip Morris are listed frequently because of their wide ownership and institutional backing. The investor should be alert to repetition. If an industry or one company suddenly appears within a short period of time, there is likely something happening as evidenced by institutional interest, however.

A Word of Caution: A stock may be very active one day for a special reason such as a possible acquisition. However, that rumored acquisition may not materialize. Further, after that one day spurt in activity, the stock may become dormant.

Also See: Trading Volume.

Tool #72

Moving Averages

What Is This Tool? A moving average (MA) is an average that is updated as new investment information is received. The investor uses the most recent stock price and/or volume to calculate an average, which is used to predict future market prices and/or volume. A moving average, also, can be used to evaluate intermediate- and long-term stock movements. The moving average shows the underlying direction and magnitude of change of very volatile numbers.

How Is It Computed? The most recent observation is used to calculate a moving average. Moving averages are constantly updated. It is determined by averaging a portion of the series and then adding the subsequent number to the numbers already averaged, omitting the first number, and obtaining a new average.

For instance, a 30-week MA records the average closing price of a stock for the 30 most recent Fridays. Each week, the total changes because of the addition of the latest week's closing figures and the subtraction of those of 30 weeks ago, then the new total is divided by 30 to obtain the MA.

Example 1: Month-end stock prices for the following months:

January	$20
February	$23
March	$20
April	$21
May	$16

The 4-month moving average for June is computed as follows.

$$\frac{23 + 20 + 21 + 16}{4} = \frac{80}{4}$$

$$= 20$$

Example 2

Day	Index	3-Day Moving Total	3-Day Moving Average
1	121		
2	130		
3	106	357 (Days 1–3)	119 (357/3)
4	112	348 (Days 2–4)	116 (348/3)
5	100	318 (Days 3–5)	106 (318/3)

Where Is It Found? Moving average information and charts may be found in brokerage research reports. Many technical analysts prepare and chart 200-day moving averages in their reports. Otherwise, the investor can determine the moving average from stock price quotations and volume figures on a stock published in financial newspapers, brokerage reports, and such. Typically, daily or weekly price changes are graphed. Market indexes are also published in financial publications such as *The Wall Street Journal.*

How Is It Used and Applied? Many analysts are of the opinion that a reversal in a significant up trend in the price of a stock and/or overall market may be identified beforehand or at least confirmed by examining the movement of current prices compared to the long-term moving average of prices. A moving average shows the direction and degree of change of a fluctuating series of prices.

How Is It Used for Investment Decisions? Moving average is used as a prediction model to determine the future expected market price of stock. The investor can choose the number of periods to use on the basis of the relative importance he or she attaches to old vs. current data. For example, an investor might compare two possibilities, a five-month and three-month period. In terms of the relative importance of new vs. old data, the old data received a weight of 4/5 and current data 1/5. In the second possibility, the old data received a weight of 2/3, while current observation received 1/3 weight.

By examining the movement of present prices compared to the long-term moving average of prices, the investor can foresee a reversal in a major up trend in price of a particular security or the general market.

A 200-day moving average of daily ending prices is usually used. The investor can graph the average on stock price charts to see directions. The investor should buy when the 200-day average line becomes constant or rises after a decline and when the daily price of stock moves up above the average line. A buy also is indicated when the stock price rises above the 200-day line, then goes down toward it but not through it, and then goes up again. The investor should consider selling when the average line becomes constant or slides down after a rise and when the daily stock price goes down through the average line. A sell also is indicated when the stock price is below the average line, then rises toward it, but instead of going through it, the price slips down again.

Tool #73

Mutual Fund Cash-to-Assets Ratio

What Is This Tool?   This ratio refers to the level of cash investments held by stock mutual funds, as compiled by Investment Company Institute, a trade group also called ICI. This is seen as an indicator of how much money leading money managers are willing to commit to the stock market.

How Is It Computed?   Each month, the ICI surveys fund companies about the assets they hold. This ratio is determined by totaling the cash holdings of stock mutual funds and dividing it by the total assets in those funds. A report is issued typically near the end of the month with statistics on the previous month's fund activity.

Where Is It Found?   The monthly ICI report is covered in such publications as *The New York Times* and *The Wall Street Journal. Barron's* runs a chart weekly on the most recent monthly data.

How Is It Used for Investment Decisions?   The ratio is seen as a measure of the stock fund manager's outlook for stocks.

At its simplest, when the level of cash is falling, it is often seen as a signal of market strength. Mutual funds represent one of the biggest stock buyers today. Conversely, when the ratio is rising, it may show that stock fund managers are getting nervous about holding stocks and that an investor may want to do likewise and lighten his exposure to equities.

Conversely, market conditions also can be brought into the analysis.

For example, when stock fund managers' exuberance for stocks brings the ratios to less than 5 percent and the market is still rising, some analysts believe that this is a signal to sell. Such analysts fear that there will be little fund buying power left to support the stock market as mutual funds become fully invested.

On the other hand, when fund managers are aggressively adding to cash in a down market, some may consider such activity a signal that the market bottom may be approaching. Indeed, in such a situation, the turnaround could be dramatic as the funds, heavy with cash, rush to buy shares once they start an upswing.

A Word of Caution: The data is dated, by at least one month, which can make analysis difficult. Also, thanks to the popularity of 401 (k) savings plans where workers constantly add new money to their accounts, many fund managers get a constant supply of fresh cash to invest. This can allow some fund managers to be steady buyers of stocks while carrying a low cash ratio. In addition, thanks to new, derivative investments that mimic stock market indexes, fund mangers can further lower their cash balances by using these derivatives as highly liquid but stock market-earning cash equivalents.

Also See: Economic Indicators and the Securities Market.

Tool #74

Mutual Funds: Evaluation Tools

ALPHA FOR A MUTUAL FUND

What Is This Tool?   Alpha is the excess return that the portfolio manager is able to earn above an unmanaged portfolio (or market portfolio) that has the same risk. In the context of a mutual fund, an alpha value is the value representing the difference between the return on a fund and a point on the market line, where the market line describes the relationship between excess returns and the portfolio beta.

How Is It Computed?   Alpha is beta × (market return − risk-free return).

Example 1:   If the market return is 8 percent and the risk-free rate (such as a rate on a T-bill) is 5 percent, the market excess return equals 3 percent. A portfolio with a beta of 1 should expect to earn the market rate of excess returns, or alpha, equal to 3 percent (1 × 3 percent). A fund with a beta of 1.5 should provide excess returns of 4.5 percent (1.5 × 3 percent).

Where Is It Found?   Morningstar Inc. (800-735-0700) offers numerous examples that show alpha values of major funds as does its Web site at www.morningstar.net.

How Is It Used and Applied?   Alpha value is used to evaluate the performance of mutual funds. Generally, a positive alpha (excess return) indicates superior performance while a negative value leads to the opposite conclusion.

Example 2:   The fund in Example 1 has a beta of 1.5, which indicates an expected excess return of 4.5 percent along the market line. Assume that the fund had an actual excess return of only 4.1 percent. That means the fund has a negative alpha of 0.4 percent (4.1 percent − 4.5 percent). The fund's performance is, therefore, inferior to that of the market.

How Is It Used for Investment Decisions?   "Keep your alpha high and your beta low" is a basic strategy for those who wish to invest in a mutual fund.

This measure should cover at least three years of data to give the most accurate picture about the performance of the fund. The key question for investors is, can a fund consistently perform at positive alpha levels?

Word of Caution: Alpha should be analyzed together with risk measures such as beta, R^2, and standard deviation and other fund selection criteria, such as fees, investment objectives, shareholder services, and the fund manager's experience.

Also See: Jensen's Performance Measure, Beta for a Mutual Fund, Mutual Funds: R^2, Mutual Funds: Standard Deviation, Sharpe's Risk-Adjusted Return.

EXPENSE RATIOS

What Are These Tools?   An expense ratio is a measure of how much it costs to own a mutual fund. Other than commissions or other sales fees, these costs are not directly billed to the investors. That makes them hard to follow. These expenses, from the fund's legal bills to the manager's profit, are taken out of the fund's net asset value.

How Are They Computed?   The ratio reflects the various expenses charged against the value of the assets in a fund. These expenses include management fees paid to the fund company, the cost of running the fund, and added charges for marketing the fund, which are called 12b-1 fees. They are totaled and then divided by the total assets in the fund. The ratio equates to the amount the fund's total return is reduced in comparison to directly owning the same securities.

The expense ratio does not include sales commissions or the fund's costs of trading securities.

Where Are They Found?   Investors can track these expenses in two places. Every mutual fund prospectus details these expenses both as a percentage of the fund's assets and expressed in terms of how much a saver is charged to own the fund over five years. In addition, many fund-watching services, most notably Morningstar of Chicago, track such charges.

How Are They Used for Investment Decisions?   Experts say that while no investment should be made solely because of a fund's level of expenses, it is something to watch closely. It is particularly important for bond funds where expenses, rather than management expertise, can be crucial. (See Figure 70.)

Also See: Lipper Mutual Fund Rankings, Morningstar Mutual Fund Rankings.

MUTUAL FUND NET ASSET VALUE

What Is This Tool?   The value of a mutual fund share is measured by *net asset value* (NAV), which tells you what each share of your mutual fund is worth.

How Is It Computed?

$$NAV = \frac{\text{Fund's total assets} - \text{liabilities}}{\text{Number of shares outstanding in the fund}}$$

For example, assume that a fund owns 100 shares each of General Motors (GM), Xerox, and International Business Machines (IBM). Assume, also, that on a particular day, the following market values existed. Then, the NAV of the fund is calculated (assume the fund has no liabilities) as

1. GM—$90 per share times 100 shares equal to $9,000.
2. Xerox—$100 per share times 100 shares equal to $10,000.
3. IBM—$160 per share times 100 shares equal to $16,000.
4. The value of the fund's portfolio is equal to $35,000.
5. The number of shares outstanding in the fund is 1,000.
6. The net asset value (NAV) per share is equal to point (3) divided by point (4), or $35,000/1,000 or $35.

Mutual Fund Expense Ratios

Fund Sector	Expense Ratio
Aggressive Growth Stocks	1.64%
Asset Allocation	1.31%
Balanced	1.38%
Corporate Bond	0.95%
Equity Income	1.32%
Foreign Stock	1.71%
Government Bond	1.12%
Growth and Income Stock	1.44%
Growth Stocks	1.23%
Junk Bond	1.33%
Money-Market funds	0.60%
Multi-Asset Global	1.90%
Multi-Sector Bond	1.53%
Muni Bond	1.03%
Small Company Stock	1.50%
Specialty Stock	2.43%
World Bond	1.42%

FIGURE 70   To evaluate a fund, an investor should look at his fund's total annual expenses and then compare his or her costs with the above industrywide averages as of September 1998. (From Morningstar Inc. With permission.)

If an investor owns 5 percent of the fund's outstanding shares, or 50 shares (5 percent times 1000 shares), then the value of the investment is $1750 ($35 times 50).

Where Is It Found?   NAV is reported in the mutual fund section of the financial pages of every daily newspaper. NAVs on mutual funds also are widely available on the Internet.

How Is It Used for Investment Decisions?   NAV represents one component of the return on mutual fund investments. There are two ways to make money in mutual funds.

1. NAV only indicates the current market value of the underlying portfolio.
2. An investor also receives capital gains and dividends. Therefore, the performance of a mutual fund must be judged on the basis of these three returns.

An investor should monitor the closing daily change in NAV of a fund. Such information provides an indicator of the return you have earned on your money. Of course, fund managers make every effort to increase the NAV, because their performance is partly evaluated on its change.

A Word of Caution: It is important to remember that there are other measures of quality of a fund such as beta, alpha, R^2, standard deviation, and risk-adjusted return.

Also See: Alpha for a Mutual Fund, Beta for a Mutual Fund; Risk-Adjusted Return; Standard Deviation for a Mutual Fund; R^2 for a Mutual Fund.

R^2 FOR A MUTUAL FUND

What Is This Tool? R^2 is the percentage of a fund's movement that can be explained by changes in the S&P 500. In statistics, it is called the *coefficient of determination*, designated R^2 (read as R-squared).

Simply put, R^2 tells us how good the overall relationship is between the dependent variable y and the explanatory variable x. More specifically, the coefficient of determination represents the proportion of the total variation in y that is explained by x. It has the range of values between 0 and 1.

How Is It Computed?   The coefficient of determination is computed as

$$R^2 = 1 - \frac{\Sigma(y - y')^2}{\Sigma(y - \bar{y})^2}$$

However, there is a short-cut method available.

$$R^2 = \frac{[n\Sigma xy - (\Sigma x)(\Sigma y)]^2}{[n\Sigma x^2 - (\Sigma x)^2][n\Sigma y^2 - (\Sigma y)^2]}$$

Example: To illustrate the computation of R^2, we will refer to the data given below:

S&P Returns (x)	XYZ Fund Returns (y)	xy	x^2	y^2
9	15	135	81	225
19	20	380	361	400
11	14	154	121	196
14	16	224	196	256
23	25	575	529	625
12	20	240	144	400
12	20	240	144	400
22	23	506	484	529
7	14	98	49	196
13	22	286	169	484
15	18	270	225	324
<u>17</u>	<u>18</u>	<u>306</u>	<u>289</u>	<u>324</u>
174	225	3414	2792	4359

From the preceding table,

$$\Sigma x = 174 \qquad \Sigma y = 225 \qquad \Sigma xy = 3414 \qquad \Sigma x^2 = 2792$$
$$\bar{x} = \Sigma x/n = 174/12 = 14.5 \qquad \bar{y} = \Sigma y/n = 225/12 = 18.75$$

Using the shortcut method for R^2,

$$R^2 = \frac{(1,818)^2}{(3,228)(52,308 - 50,625)} = \frac{3,305,124}{(3,228)[(12)(4,359) - (225)^2]}$$

$$= \frac{3,305,124}{(3,228)(1,683)} = \frac{3,305,124}{5,432,724} = 0.6084 = 60.84\%$$

Excel Regression output shows the following.

SUMMARY OUTPUT

Regression Statistics

Multiple R	0.77998286
R^2	0.60837326
Adjusted R^2	0.56921058
Standard Error	2.34362221
Observations	12

This means that about 60.84 percent of the total variation in XYZ Fund returns is explained by the market (represented in this example by the S&P 500 Index). A relatively low R^2 indicates that there is little correlation between XYZ Fund and the market.

Where Is It Found? Morningstar Inc. (800-876-5005) offers numerous studies that show alpha values of major funds as does its web site at www.morningstar.net. (See Figure 71.)

How Is It Used for Investment Decisions? If the R^2 of a mutual fund is close to 100 percent, the fund is well diversified. The further away a mutual fund is from 100 percent, the less the fund is diversified.

A Word of Caution: This measure should cover at least three years of data to give the most accurate picture about the performance of the fund.

Further, R^2 should be analyzed together with risk measures such as beta and standard deviation and other fund selection criteria, such as fees, investment objectives, shareholder services, and the fund manager's experience.

Also See: Alpha for a Mutual Fund, Beta for a Mutual Fund, Risk-Adjusted Return, Standard Deviation for a Mutual Fund.

STANDARD DEVIATION FOR A MUTUAL FUND

What Is This Tool? The standard deviation is the measure of the tightness of the probability distribution. In other words, it measures the tendency of data to be spread out. Investors can make important inferences from past data with this measure. The standard deviation, denoted with the Greek letter σ, read as sigma, is defined as follows.

$$\sigma = \sqrt{\frac{(x - \bar{x})^2}{n}}$$

where x is the mean (arithmetic average).

How Is It Computed?   More specifically, the standard deviation can be calculated, step-by-step, as follows:

1. Subtract the mean from each element of the data.
2. Square each of the differences obtained in Step 1.
3. Add together all the squared differences.
4. Divide the sum of all the squared differences by the number of values minus one.
5. Take the square root of the quotient obtained in Step 4.

Example: Quarterly returns are listed for next six quarters for ABC Mutual Fund.

Time period	x	$(x - \bar{x})$	$(x - \bar{x})^2$
1	10%	0	0
2	15	5	25
3	20	10	100
4	5	−5	25
5	−10	−20	400
6	20	10	100
	60		650

From the preceding table, note that

$$\bar{x} = 60/6 = 10 \text{ percent}$$

$$\sigma = \sqrt{\frac{(x - \bar{x})^2}{n}} = \sqrt{\frac{650}{6}} = \sqrt{108} = 10.41 \text{ percent}$$

ABC Fund has returned on the average 10 percent over the last six quarters and the variability about its average return was 10.41 percent. The high standard deviation (10.41 percent) relative to the average return of 10 percent indicates that the fund is very risky.

Where Is It Found?   Morningstar Inc. (800-735-0700) offers numerous studies that show alpha values of major funds as does its Web site at www.morningstar.net.

How Is It Used and Applied?   The calculation of the standard deviation may not seem simple, but the calculation can be done easily with a calculator with a square root key, a financial calculator with the standard deviation key, or spreadsheet programs such as Microsoft Excel that have a built-in standard deviation function.

The standard deviation can be used to measure the variation of such items as the expected return. It can also be used to assess the risk associated with investments.

It is used to measure risk for a security. The smaller the standard deviation, the tighter the probability distribution, and, accordingly, the lower the riskiness of the security.

How Is It Used for Investment Decisions?   If the expected rate of return of XYZ Fund is 15 percent and the standard deviation is 10 percent, the actual return of the fund will be in the following range.

1. Probability of 68 percent.

The actual return is equal to 15 percent plus or minus 10 percent, from 5 to 25 percent.

2. Probability of 95 percent.

The actual return is equal to 15 percent plus or minus 20 percent, from −5 to +35 percent. It is assumed the probability distribution of returns is normal.

A Word of Caution: In calculating the standard deviation, there are a number of related issues, such as the divisor used, or the minimum number of data points necessary for the calculation to function properly. Generally speaking, the fewer data points used, the less useful the calculation of volatility. To properly measure the volatility (variability) of the investment's return, you must also consider other measures such as beta.

Also See: Alpha for a Mutual Fund, Beta for a Mutual Fund, Risk-Adjusted Return, R^2 for a Mutual Fund.

Mutual Fund Analytical Tools

Fund Name	R^2	Alpha	Beta	Standard Deviation	Sharpe Ratio	3-Year Total Return
Aggressive Growth Stocks	68%	−14.1%	1.18	24.9%	0.09	7.2%
Asset Allocation	82%	−2.5%	0.58	10.8%	0.73	12.0%
Balanced	86%	−2.6%	0.60	11.1%	0.77	12.6%
Corporate Bond	86%	−0.6%	0.88	3.7%	0.82	7.6%
Equity Income	86%	−2.3%	0.78	14.9%	0.83	16.0%
Foreign Stock	51%	−11.7%	0.74	17.0%	−0.10	3.6%
Government Bond	90%	−0.9%	0.86	3.4%	0.64	7.2%
Growth and Income Stock	81%	−7.0%	1.02	20.3%	0.51	14.0%
Growth Stocks	90%	−3.2%	0.92	17.5%	0.78	17.2%
Junk Bond	2%	2.8%	0.06	6.7%	0.54	8.2%
Multi-Asset Global	72%	−5.6%	0.46	11.6%	0.19	6.7%
Multi-Sector Bond	15%	0.3%	0.38	5.5%	0.41	6.7%
Muni Bond	68%	−0.7%	0.70	3.2%	0.46	6.9%
Small Company Stock	57%	−12.4%	1.03	23.2%	0.10	6.8%
Specialty Stock	63%	−7.5%	0.90	20.0%	0.34	11.2%
World Bond	23%	0.2%	0.52	8.8%	0.35	6.6%

FIGURE 71   Here is a look at some key mutual fund categories and how they compare in some important analytical tools: R^2 (three years of data), alpha (three years), beta (three years); standard deviation (three years), Sharpe ratio, and total return (three-year annual average). (From Morningstar Inc. With permission.)

TOOL #75

MUTUAL FUNDS: SALES FIGURES

What Is This Tool?   The fund industry's trade group issues a monthly report on sales and redemption of stock, bond, and money market funds. The report is usually out four weeks after a month ends. The report is seen as a reflection of small investors' opinion on the climate for buying stocks or bonds or for holding cash. Several independent newsletters also track this data on a monthly and weekly basis.

How Is It Computed?   The Investment Company Institute (ICI) polls its members on their cash inflows and outflows. ICI then tabulates industrywide figures, excluding such items as dividend payments. The report includes figures on new sales, redemptions, and the most closely watched net sales figure. Two newsletters, *Trimtabs.com* and *AMG Data Services,* compile weekly estimates of fund inflows and outflows by talking to some major fund families about their sales patterns.

Where Is It Found?   Monthly sales figure are grist for regular news stories in daily newspapers such as *The Wall Street Journal* and *The New York Times.* The newsletter data has gotten wide coverage, too. The ICI release and other statistical releases are available at the ICI's Web site at www.ici.org. You can check out fund-sales tracking newsletters on the Internet at trimtabs.com or www.amgdata.com.

How Is It Used for Investment Decisions?   With the power that mutual funds possess, the monthly reports are seen as good indicators of the individual investor's appetite for stocks and/or bonds. There is some debate, though, over the meaning of small investors' buying habits. Some experts feel these savers tend to be bad investors and should be considered a contra-indicator. Others, however, say the small investor is a powerful market force, one that can lead stock or bond advances or declines.

In addition, the monthly ICI report contains a reading on the cash position of stock funds. This figure is watched as an indicator of the bullish or bearish nature of mutual fund managers.

A Word of Caution: Fund investors, much like the consuming public, can have buying habits altered by marketing campaigns. In some cases, fund purchases or redemptions may reflect fund companies' marketing strategies, altering the industrywide numbers' ability to reflect investor sentiment. The ICI figures, while comprehensive, are dated. The newsletter figures are more timely but are compiled from just a small slice of the industry and can give misleading signals.

Also See: Mutual Fund Cash-to-Assets Ratio.

TOOL #76

NASDAQ INDEXES

What Are These Tools?   The NASDAQ indexes follow the price performance of over-the-counter securities. The indexes are shown in Figure 72.

- *NASDAQ Composite*—a widely quoted index that measures all NASDAQ domestic and non-U.S. based common stocks listed on the NASDAQ Stock Market. The Composite includes over 5000 companies, more than most other stock market indexes. Because NASDAQ today trades such major issues as Microsoft, Intel, and Cisco Systems, this index no longer is considered the barometers of small stocks that it once was.
- *NASDAQ National Market Composite*—a subset of the NASDAQ Composite comprising all companies included in the NASDAQ Composite listed on the top tier of the NASDAQ Stock Market.
- *NASDAQ 100*—includes 100 of the largest nonfinancial domestic companies listed on the NASDAQ National Market tier. Launched in January 1985, NASDAQ 100 reflects NASDAQ's largest growth companies across major industry groups. It is also the basis for several stock index option contracts.
- *NASDAQ Financial 100*—consists of 100 of the largest financial organizations listed on the NASDAQ National Market tier.
- *NASDAQ Bank Index*—includes more than 350 banking companies including trust companies and firms that support banking functions such as check cashing agencies, currency exchanges, safe deposit companies, and banking corporations overseas.
- *NASDAQ Biotechnology Index*—includes about 100 companies engaged in biomedical research to develop new treatments and cures for disease.
- *NASDAQ Computer Index*—includes more than 500 computer hardware and software companies, data processing services, and firms that produce office equipment and electronic components.
- *NASDAQ Insurance Index*—includes about 100 insurance companies including life, health, property, casualty, brokers, agents, and related services.
- *NASDAQ Other Finance Index*—some 600 credit agencies (except banks), savings and loans, security and commodity brokers, real estate firms, and others in financial services.
- *NASDAQ Transportation Index*—includes about 90 railroads, trucking companies, airlines, pipelines (except natural gas), and services such as housing and travel arrangements.

U.S. stock indexes

Index	High	Low	Close	Net Chg.
Dow Jones Industrials	11310.55	11104.86	11215.10	+112.09
Dow Jones Transportation	2740.66	2711.00	2723.63	-2.54
Dow Jones Utilities	365.66	359.51	363.74	+3.62
Dow Jones Composite	3272.62	3223.98	3249.83	+26.03
NYSE Composite	678.22	670.40	674.53	+4.07
NYSE Industrials	848.79	842.02	843.87	+1.61
NYSE Transportation	406.99	402.49	404.89	+1.46
NYSE Utilities	470.22	462.78	467.53	+4.75
NYSE Finance	617.50	603.41	614.43	+11.02
AMEX Composite	946.13	934.22	943.48	+9.26
Nasdaq 100	4080.48	3971.32	4077.59	+108.86
Nasdaq Composite	4208.73	4127.19	4206.35	+102.54
Nasdaq Industrials	2266.27	2221.25	2265.16	+53.49
Nasdaq Banks	1680.67	1651.28	1673.94	+23.09
Nasdaq Insurance	1900.84	1868.88	1898.47	+25.31
Nasdaq Financial	2904.74	2876.47	2901.49	+15.20
Nasdaq Transportation	1115.82	1103.00	1115.65	+8.30
Nasdaq Telecom	831.80	817.04	831.52	+20.21
Nasdaq NMS Comp	1916.36	1879.18	1915.25	+46.75
Nasdaq NMS Indust	935.49	916.80	935.00	+22.13
Nasdaq Computer	2532.18	2481.67	2529.99	+62.78
Nasdaq Biotech	1372.03	1323.20	1362.95	+46.72
S&P 100	832.18	820.93	827.41	+6.48
S&P 500	1525.30	1502.59	1517.68	+15.09
S&P Midcap	543.97	535.80	542.90	+7.10
S&P Industrials	1854.16	1830.29	1844.92	+14.63
S&P Transportation	603.81	594.70	598.74	+1.41
S&P Utilities	311.49	306.84	310.24	+2.89
S&P Smallcap	223.92	221.37	223.49	+2.12
S&P Financial	158.61	154.31	157.73	+3.42
Major Market Index	1079.93	1069.26	1069.26	-1.92
Value Line Arith	1163.51	1151.75	1160.44	+8.69
Russell 2000	539.14	532.36	537.89	+5.56
Wilshire Smallcap	879.75	862.72	874.69	+11.97
Value Line Geometric	435.39	435.39	435.39	+3.03
Wilshire 5000	14280.04	14280.04	14280.04	+154.46

FIGURE 72   How a newspaper tracks NASDAQ stock indexes. (From *Orange County Register* (CA), September 1, 2000, p. B6. With permission.)

- *NASDAQ Telecommunications Index*—about 150 telecommunications companies.
- *NASDAQ Industrial Index*—about 3100 companies involved in agricultural, mining, construction, manufacturing, services, and public administration.

How Are They Computed?   The Index is market-value weighted so each company's security affects the Index in proportion to its market value.

Where Are They Found?   Most daily newspapers, business television shows, and on-line computer services track some of these indexes, especially the NASDAQ Composite. The NASDAQ Web site at www.nasdaq.com has up-to-the-minute quotes on the indexes throughout the trading day.

How Are They Used for Investment Decisions?   Moves in NASDAQ stocks are seen as an indication of how second-tier issues are doing. The NASDAQ Composite, laden with numerous leaders in technology fields, is seen as a harbinger of how technology stocks are faring.

If an index is very low, an investor may expect an increase to more normal levels. A low index may represent a buying opportunity in selected stocks. An overbought situation is when the index is at its peak and the investor believes that a correction is imminent. In this case, the investor may sell overpriced stocks he or she owns or sell short in expectation of a downturn in price.

A Word of Caution: The NASDAQ Stock market tends to be home to numerous volatile issues from young companies. That means that the performance of any NASDAQ index, especially the subindexes that contain a smaller number of companies, may be a poor indicator of how individual issues are faring.

Also See: AMEX Indexes, Russell 2000 Index.

Tool #77

New Highs-to-Lows Ratio

What Is This Tool?   The new highs–new lows ratio is the number of new issues that traded at their highest 52-week price divided by the number of issues that traded at their lowest 52-week price. It is typically tracked for shares trading on the New York Stock Exchange, but can also be calculated for the NASDAQ and American markets.

How Is It Computed?   The new highs–new lows ratio equals

$$\frac{\text{Number of issues at 52-week highest price}}{\text{Number of issues at 52-week lowest price}}$$

Example   New highs are 100. New lows are 50.

$$\frac{20}{100} = 0.20$$

The ratio is therefore 0.20, which is low. This is a bearish situation.

Where Is It Found?   Listings of new highs and new lows for the trading day are published in the financial pages of newspapers and magazines such as *Barron's, Investor's Business Daily, The New York Times,* and *The Wall Street Journal.* (See Figure 73.)

How Is It Used for Investment Decisions?   An uptrend in the ratio of new highs to new lows is a bullish indicator of the stock market while a downtrend is a bearish sign.

The investor may look at specific companies having new highs as the basis to make an investment decision. If the investor feels the stock is overvalued, he or she should not buy it or, if it is currently held, sell it. Of course, if the investor believes the company will do even better, the stock should be retained. On the other hand, a company listed as a new low might be bought if the investor feels the price is overly depressed and the company has potential.

A Word of Caution: The number of new highs to new lows is just one measure of performance. Other indicators of stock performance must be examined as well. For example, there might be more new lows than new highs but the Standard & Poor's 500 index may have increased. That raises questions about the broad-based nature of such a rally.

Also See: Breadth (Advance–Decline) Index, New Highs and Lows: New York Stock Exchange High–Low Index.

WEEK'S NEW HIGHS AND LOWS			
Weekly Comp.	NYSE	AMEX	Nasdaq
New Highs	254	65	214
New Lows	116	63	267

FIGURE 73 How a newspaper covers new highs and new lows. (From *Barron's Market Week,* August 28, 2000, p. MW74. With permission.)

Tool #78

New York Stock Exchange Indexes

What Are These Tools? In 1966, the New York Stock Exchange started indexes consisting of the Composite Index and its four subindexes which are the Industrial Index, Transportation Index, Utilities Index, and Financial Index.

How Are They Computed? The Composite Index is a capitalized market value-weighted index of all NYSE issues. The indexes measure changes in aggregate market value of NYSE common stocks, adjusted for new issues, mergers, bankruptcies, and stocks splits. Any change in an individual issue requires a proportionate change in the market value of the base figure.

The index is computed as follows.

1. Multiply the market value of each common share by the number of shares.
2. Add the results for all issues to derive the total market value.
3. The index is a number showing the relationship between total current market value and a base market value (established in 1966). Any needed adjustments are made. (50 is the base value for the index.) Point changes are in terms of dollars.

Example: For the year-end total market value of common stocks, $5943.5 billion; adjusted base market value, $901.9 billion,

$$\frac{\$5943.5}{901.9 \times 50} = 329.51 \text{ at year-end}$$

The Industrial Index is comprised of more than 1000 industrial companies. The Transportation Index is comprised of transportation companies such as air carriers, truckers, and railroads. The Utilities Index tracks about 40 utilities including the areas of electric power, gas, water, and telecommunications. The Financial Index tracks about 400 financial companies including the areas of banking, insurance, credit, brokerage, and investment.

Where Are They Found? The NYSE indexes are reported in most daily newspapers, business television shows, and through online computerized investment databases. The indexes can also be found the NYSE Web site at www.nyse.com. (See Figure 74.)

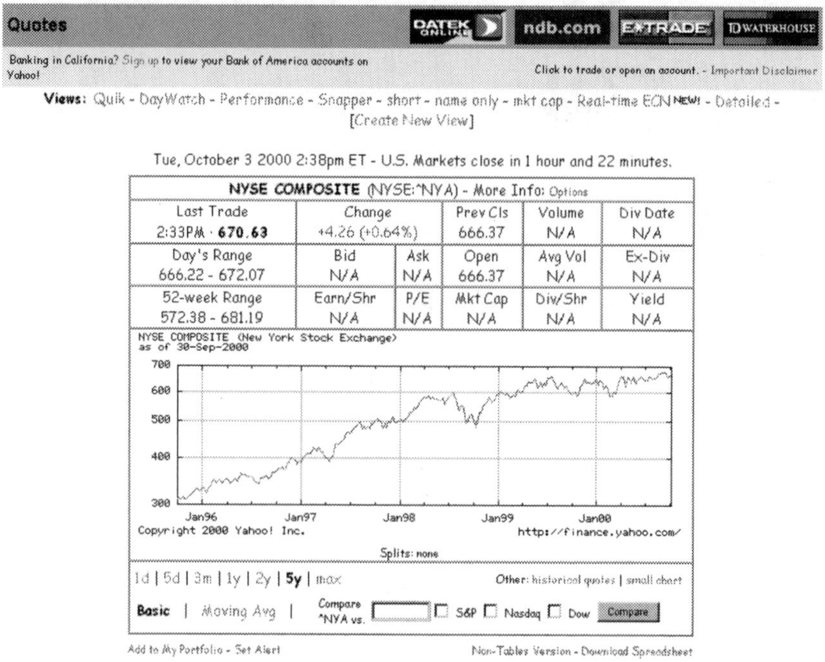

FIGURE 74   How a web site covers the NYSE indexes. (Reproduced with permission of Yahoo! Inc.© 2000 by Yahoo! Inc. YAHOO! and the YAHOO! Logo are trademarks of Yahoo! Inc.)

How Are They Used for Investment Decisions?   The New York Stock Exchange Composite Index is a measure of the performance of New York Stock Exchange issues, considered a home to major companies. So these indexes are often seen as a barometer of market conditions for large, blue-chip issues. If the index consistently increases, a bull market exists.

However, if the index consistently decreases, there is a bear market. Therefore, investors can use the index as one indicator of whether to buy stocks or to avoid them. A healthy, expanding stock market with future upward potential provides a buying opportunity.

The New York Stock Exchange Composite Index may be looked at as confirmation of a change in other indexes such as the Dow Jones Industrial Average and Standard & Poor's 500. If all three indexes move in the same direction with similar magnitude, there is consistency. Thus, the investor can be more confident of the degree of market strength or weakness when several indexes confirm each other.

The NYSE Composite Index also provides the base for options written on the Index and for futures contracts.

Tool #79

Nikkei (Tokyo) Stock Index

What Is This Tool?   It is the most widely watched barometer of Japan's stock market, the world's second largest behind the United States. The Index, first published in May 1949, is to Japanese shares what the Dow Jones Industrial Average is to U.S. issues.

How Is It Calculated?   The Nikkei is an average of prices of 225 stocks listed in the prestigious First Section of the Tokyo Stock Exchange. Companies included are Asahi Breweries, Fuji Film, Nippon Steel, and Yamaha.

Where Is It Found?   Index information appears in major local and national newspapers and business dailies. Thanks to the international date line, *USA Today* can put the afternoon Nikkei index results on the front page of its morning paper. This index can also be tracked by online databases like America Online. (See Figure 75.)

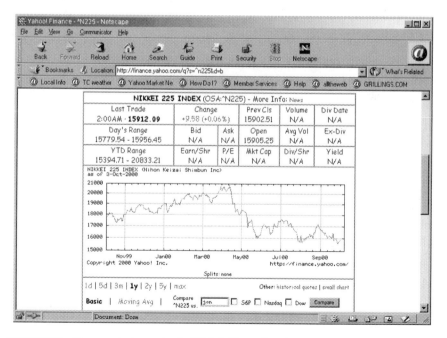

FIGURE 75   How a web site tracks the Nikkei. (Reproduced with permission of Yahoo! Inc.© 2000 by Yahoo! Inc. YAHOO! and the YAHOO! Logo are trademarks of Yahoo! Inc.)

How Is It Used for Investment Decisions?   The Index tracks movement in Japanese share prices. Sharp gains in the Nikkei, especially in relationship to U.S. market indexes, may be a signal to move money across the Pacific to Tokyo. Conversely, a steep drop in the Nikkei, it has never regained its 1989 highs, can be a warning sign that Japanese markets are quite volatile and are worth avoiding.

A Word of Caution: The Nikkei may be best known for its five-digit numerology which to the uninformed eye makes an already temperamental market seem very unnerving. Remember to keep that in perspective.

Also, some experts feel that the Nikkei is too narrow a measure of Japanese stocks. The Tokyo Stock Exchange Index, or Topix, covers all of approximately 1200 shares in the First Section and is viewed as a better barometer of Japanese market conditions.

Fluctuations can play a major part of any overseas investment. (See Figure 76.)

Foreign Stock Indexes

- *All-Ordinaries Index*—tracks all shares listed on the Australian stock exchanges.
- *Banca Commerciale Italiana General Index*—includes over 325 Italian companies on Milan Exchange.
- *Bangkok Book Club Index*—index of all the shares listed on the Securities Exchange of Thailand.
- *Bombay SE Sensitive Index*—"SENSEX" is a capitalization-weighted price index which includes the 30 leading Indian stocks based on market capitalization and activity.
- *Bovespa Index*—includes stocks which represent about 85 percent of all transactions on Brazilian exchanges. Stocks are weighted by their participation in share trading.
- *CAC 40 Index*—capitalization-weighted price index of the 40 most actively French traded shares on the Paris stock exchange. CAC stands for Compagnie des Agents de Change.
- *Commerzbank Index*—oldest daily German share index based on 78 blue chip stocks which represent about 75 percent of the country's total market value.
- *FT-Actuaries World Nordic Index*—*The Financial Times* of London's market value-weighted barometer of over 90 company issues traded in four Scandinavian markets.
- *FT-All Shares Index*—*The Financial Times* of London's market value-weighted barometer of over 750 company issues traded in the U.K. market.
- *Hang Seng Index*—market value-weighted barometer of the company issues of the 33 largest companies in the Hong Kong market.
- *IBEX 35 Index*—index of the 35 most traded stocks of the four Spanish exchanges.
- *IPC (Indice de Precios y Cotizaciones)*—capitalization-weighted price index which includes the 40 representative Mexican stocks based upon trading value and volume as well as capitalization.
- *JSE Gold Index*—index tracking leading mining companies on the Johannesburg Exchange in South Africa.
- *Korea Composite Index*—tracks all stocks listed on the Seoul Exchange in South Korea.
- *Maof Index*—covers the 25 largest Israeli shares on the Tel Aviv Exchange.
- *Oslo SE Total Index*—yield Index which includes dividend payments for all Norwegian companies on Olso Exchange's main list.
- *SBF 250*—the Societe des Bourses Francaises (SBF) 250 is a market value-weighted barometer of the company issues of the 250 largest companies in the French market.
- *Second Section Index*—market value-weighted barometer that reflects the performance of the smaller, less-established and newly listed companies of the Tokyo Exchange.
- *Shanghai "B" Shares Index*—tracks Chinese shares which can be purchased by foreigners and which are traded on the Shanghai Exchange. "A" shares can only be purchased by Chinese citizens.
- *TOPIX*—the Tokyo Stock Exchange Index (TOPIX) is a market value-weighted barometer of over 1100 company issues traded in the Japanese market.
- *TSE 300*—the Toronto Stock Exchange (TSE) 300 is a market value-weighted barometer of 300 company issues traded in the Canadian market.

FIGURE 76　Some other widely watched foreign indexes.

Tool #**80**

Option Tools

CBOE PUT–CALL RATIO

What Is This Tool?   The Chicago Board Options Exchange (CBOE) ratio compares put volume to call volume on option contracts. In a put, the investor bets the market will go down. In a call, the investor bets that the market will go up.

How Is It Computed?

$$\text{CBOE Put–Call Ratio} = \frac{\text{Put volume}}{\text{Call volume}}$$

Example: The following option volume information is presented for the week.

	Puts	Calls
S&P 100	600,000	500,000
CBOE Equity	200,000	600,000

The ratios are

$$\text{S\&P } 100 = 600,000/500,000 = 120/100$$
$$\text{CBOE Equity} = 200,000/600,000 = 33.3/100$$

Where Is It Found?   The CBOE ratio appears in *Barron's*.

How Is It Used for Investment Decisions?   Investors must examine the trend in the ratio over time (e.g., weekly). A put–call ratio of 70 puts to every 100 calls on the S&P's 100 and 65 puts to every 100 calls on the CBOE Equity ratio is a positive sign. However, a put–call ratio of only 40 puts to 100 calls on the S&P 100 and on the CBOE Equity ratio is a negative indicator. The CBOE put–call ratio is a contrarian tool. For example, a low put volume reflects bullishness.

A Word of Caution: Although the put–call ratio is expected to be a contrary indicator, this may not always be the case.

Also See: CBOE Options: Put–Call Options Premium Ratio.

PUT–CALL OPTIONS PREMIUM RATIO

What Is This Tool?   This is a short-term contrarian method to predict future movement in stock prices. A premium is the cost to the investor of buying a call or put option.

How Is It Computed?   The Put–Call Options Premium Ratio equals

Average premium on listed put options
Average premium on listed call options

A put option is the right to sell a company's shares at a given price by a certain date while a call option is the right to buy those shares. With a put option, an investor makes money if the stock price decreases; with a call option an investor makes money if the stock price increases.

Example: Assume an average premium on listed put and call options of 110 and 130, respectively. The put–call options premium ratio is 84.6 percent.

Where Is It Found?   The ratio is published in *Barron's*.

How Is It Used for Investment Decisions?   The investor should use this ratio as a technical investment analysis contrary approach. Put premiums are usually high when investors feel bearish. However, from a contrarian frame of reference, the time to buy is when investors are negative because a bullish stock market is deemed to occur. On the other hand, a low ratio means that investors are positive, which is a bearish indicator. For example, a ratio above 90 percent may indicate a stock market bottom and, therefore, a buying opportunity. A ratio below 40 percent may indicate a stock market peak and, therefore, the time to sell.

A Word of Caution: The contrarian approach does not always work.

Also See: Options: Valuation of Options (Calls and Puts).

SPREAD STRATEGY

What Is This Tool?   A spread is the purchase of an option (long position) and the writing of an option (short position) in the same security, using call options. A sophisticated investor may write many spreads to gain from the differences in option premiums. The return potential is significant, but the risk is very high. There are different types of spreads.

- A vertical spread is the purchase and writing of two contracts at different striking prices with the same expiration date.
- A horizontal spread is the purchase and writing of two options with the same strike price but for different periods.
- A diagonal spread combines the horizontal and vertical.

How Is It Computed?   Spreads require the investor to buy one call and sell another call. The gain or loss from a spread position depends on the change between two option prices as the price of the stock increases or decreases. The difference between two option prices is the price spread.

Where Is It Found?   A spread may be bought through many brokerage houses and members of the Put and Call Brokers and Dealers Association. Put and call traders devise many spread situations involving different maturities or different strike prices.

How Is It Used for Investment Decisions?   The speculator who uses a vertical bull spread anticipates an increase in the price of stock, but this strategy reduces the risk. There is a ceiling on the gain or loss.

The speculator using a vertical bear spread expects the stock price to decline. This investor sells the call short with the lower strike price and places a cap on upside risk by buying a call with a higher strike price.

Also See: Options: Put–Call Options Premium Ratio, Options: Straddling Strategy, Options: Valuation of Options (Calls and Puts).

STRADDLING STRATEGY

What Is This Tool?   Straddling integrates a put and call on the same stock with the identical exercise (strike) price and exercise date. It is employed to take advantage of significant variability in stock price. High beta stocks might be most suited for this. A straddle may be bought either to maximize return or to minimize risk. This investment approach should be left to sophisticated investors.

How Is It Computed?

$$\text{Profit on call or put } - \text{ cost of call } - \text{ cost of put } = \text{ net gain}$$

Where Is It Found?   A straddle is not traded on listed exchanges but rather must be acquired through brokerage houses and members of the Put and Call Brokers and Dealers Association.

Example: An investor buys a call and put for $4 each on September 30 when the stock price is $42. The expiration period is four months. The investment is $8, or $800 in total. Assume the stock increases to $60 at expiration of the options. The call earns a profit of $14 ($18 − $4) and the loss on the put is $4. The net gain is $10, or $1000 altogether.

How Is It Used for Investment Decisions?   A straddle is used by a speculative investor trading on both sides of the market. The speculative investor hopes for significant movement in stock price in one direction so as to make a gain that exceeds the cost of options. If the price movement does not go as expected, however, the loss will equal the cost of the options. The straddle holder may widen risk and profit potential by closing one option before closing the other.

Investors using straddles often use extensive computer analysis.

Also See: Options: Put–Call Options Premium Ratio, Options: Spread Strategy, Options: Valuation of Options (Calls and Puts).

VALUATION OF OPTIONS (CALLS AND PUTS)

What Is This Tool?   Calls and puts are a type of stock option that may be bought or sold in round lots, usually 100 shares. They come in bearer negotiable form and have a life of one to nine months.

The investor who purchases a call is buying the right to buy 100 shares of a stock at a fixed exercise price for a predetermined period. He does this when he expects the price of that stock to rise. In buying a call, the investor stands a chance of making a significant gain from a small investment, but he also risks losing his full investment if the stock does not rise in price.

Purchasing a put gives an investor the right to sell 100 shares of a stock at a fixed exercise price for a predetermined period. An investor might buy a put when he expects a stock price to fall. By purchasing a put, he gets an opportunity to make a considerable gain from a small investment, but he will lose the entire investment if the stock price does not fall.

The purchase of calls and puts gives the investor tremendous financial leverage with limited downside risk. The buying of puts can be used as a strategy for protecting unrealized capital gains.

How Is It Computed?   The cost of an option is referred to as a premium. It is the price the purchaser of the call or put has to pay the writer. (With other securities, the premium is the excess of the purchase price over a determined theoretical value.)

The premium for a call depends on

- The dividend trend of the related security.
- The volume of trading in the option.
- The exchange on which the option is listed.
- The variability in price of the related security. (A higher variability means a higher premium because of the greater speculative appeal of the option.)
- Prevailing interest rates.
- The market price of the stock to which it relates.
- The width of the spread in price of the stock relative to the option's exercise price. (A wider spread means a higher price.)
- The amount of time remaining before the option's expiration date. (The longer the period, the greater the premium's value.)

When the market price exceeds the strike price, the call is said to be in the money. However, when the market price is less than the strike price, the call is out of the money. Call options in the money have an intrinsic value equal to the difference between the market price and the strike price.

$$\text{Value of call} = (\text{market price of stock} - \text{exercise price of call}) \times 100$$

Assume that the market price of a stock is $45, with a strike price of $40. The call has a value of $500.

Out-of-the-money call options have no intrinsic value.

If the total premium (option price) of an option is $7 and the intrinsic value is $3, there is an additional premium of $4 arising from other considerations. In effect, the total premium consists of the intrinsic value plus speculative premium (time value) based on factors such as risk, variability, forecasted future prices, expiration date, leverage, and dividend.

$$\text{Total premium} = \text{intrinsic value} + \text{speculative premium}$$

The definition of in the money and out of the money are different for puts because puts permit the owner to sell stock at the strike price. When the strike price exceeds the market price of the stock, the investor has an in-the-money put option. Its value is determined as follows:

Value of put = (exercise price of put − market price of stock) × 100

Assume the market price of a stock is $53 and the strike price of the put is $60. The value of the put is $700.

When the market price of the stock exceeds strike price, there is an out-of-the-money put. Because a stock owner can sell it for a greater amount in the market than he could get by exercising the put, there is no intrinsic value of the out-of-the-money put.

	XYZ Calls at 50 Strike Price Stock price	XYZ Puts at 50 Strike Price Stock price
In the money	Over 50	Under 50
At the money	50	50
Out of the money	Under 50	Over 50

The theoretical value for calls and puts indicates the price at which the options should be traded. Typically, however, they are traded at prices higher than true value when options have a long period to go. This difference is referred to as investment premium.

$$\text{Investment premium} = \frac{\text{option premium} - \text{option value}}{\text{option value}}$$

Assume a put has a theoretical value of $1500 and a price of $1750. It is, therefore, traded at an investment premium of 16.67 percent.

Example 1:   A two-month call option allows an investor to acquire 500 shares of XYZ Co. at $20 per share. Within that time period, he exercises the option when the market price is $38. He makes a gain of $9000 before paying the brokerage commission. If the market price had declined from $20, he would not have exercised the call option, and he would have lost the cost of the option.

Example 2:   Significant percentage gains on call options are possible from the low investment compared to the price of the related common stock. Assume a stock has a present market price of $35. A call can be purchased for $300 allowing the acquisition of 100 shares at $35 each. If the price of the stock increases, the call will also be worth more. Assume that the stock is at $55 at the call's expiration date. The profit is $20 on each of the 100 shares of stock in the call, or a total of $2000 on an investment of $300. A return of 667 percent is, thus, earned. In effect, when the holder exercises the call for 100 shares at $35 each, he or she can sell them immediately at $55 per share. Note that the investor could have earned the same amount by investing directly in the common stock, but the investment would have been $3500, so the rate of return would have been significantly lower.

Example 3: An investor can buy a call giving him the right to acquire 100 shares of $30 stock at $27. The call will trade at a price of about $3 a share. Call options also may be used when an investor believes the stock price will increase in the future but has a cash flow problem and is unable to buy the stock. He or she will, however, have sufficient cash to do so later. In this situation, he or she can buy a call so as not to lose a good investment opportunity.

Example 4: On February 6 an investor purchases a $32 June call option for $3 a share. If the stock has a market price of $34 1/7, the speculative premium is $1/2. In June, he exercises the call option when the stock price is $37. The cost of the 100 shares of stock for tax reporting is the strike price ($32) plus the option premium ($3), or $35.

Example 5: A stock has a market price of $35. An investor acquires a put to sell 100 shares of stock at $35 per share. The cost of the put is $300. At the exercise date of the put, the price of the stock goes to $15 a share. He or she therefore, realizes a profit of $20 per share, or $2000. As the holder of the put, the investor simply buys on the market 100 shares at $15 each and then sells them to the writer of the put for $35 each. The net gain is $1700.

Example 6: Assume that TUV Co. stock's price was $55 on March 2. An investor buys a $56 June put for $4. The speculative premium is therefore $3. On June 7, the stock price falls to $47 and the price of the June $56 put to $8. The intrinsic value is $9 and the speculative premium is $1. As the put holder, the investor now has a gain of $4.

Example 7: As an example of a hedge, an investor buys 100 shares of JJJ Corp. at $26 each and a put for $200 on the 100 shares at an exercise price of $26. If the stock remains static, he will lose $200 on the put. If the price decreases, his or her loss on the stock will be offset by his or her gain on the put. If the stock price rises, he or she will earn a capital gain on the stock and lose the investment in the put. In other words, to get the benefit of a hedge, an investor must incur a loss on the put. (Also note that at the expiration of the put, he or she incurs a loss with no further hedge.)

Example 8: An investor buys a put to hedge his position on a stock. He or she holds 100 shares of BBC Corp. stock purchased at $60 a share. That stock increases to $80, earning a profit of $20 a share. To guarantee profit, he or she buys a put with an $80 exercise price at a cost of $300. No matter what happens later, he or she will have a minimum gain of $1700. If the stock price falls, he or she will realize an additional profit.

Example 9: An investor might buy a call to protect a short sale from the risk of increasing stock price. By doing this, the investor hedges his or her position as follows: when he or she uses a call, as a short seller this investor will not suffer a loss in excess of a given amount. However, the investor has lowered profit by the cost of the call.

Example 10: A speculator purchases an option contract to buy 100 shares at $25 a share. The option costs $150. Assume a rise in stock price to $33 a share. The speculator exercises the option and sells the shares in the market, realizing a gain of $650 ($33 − $25 − $1.50 = $6.50 × 100 shares). Now the speculator can sell the option in the market and make a profit because of its increased value. However, if there is a decline in stock price, the loss to the holder is limited to $150 (the option's cost). Of course, brokerage fees are also involved. In effect, this call option permitted the speculator to purchase 100 shares worth $2500 for $150 for a short period.

Where Is It Found?   Calls, puts, and index options are listed in financial publications such as *Barron's, Investor's Business Daily, The New York Times,* and *The Wall Street Journal.*

Calls and puts are typically written for widely held and actively traded stock on organized exchanges. Calls and puts are traded on the NYSE, AMEX, and NASDAQ markets and on listed option exchanges, which are secondary markets like the Chicago Board Options Exchange, AMEX, Philadelphia Stock Exchange, and Pacific Stock Exchange.

How Is It Used and Applied?   Options can be traded for speculative or conservative purposes. Commissions and transaction costs are involved when a call or put is purchased or sold or written. Brokerage fees depend on the amount and value of the option contract. For instance, a contract with a value ranging from $100 to $800 might have a fee of about $35.

With calls there are no voting rights, ownership interest, or dividend income. However, option contracts are adjusted for stock splits and stock dividends.

Calls and puts are not issued by the company with the common stock but rather by option writers. The maker of the option receives the price paid for the call or put minus commission costs. Calls and puts are written by, and can be acquired through, brokers and dealers. The writer is required to purchase or deliver the stock when requested.

Holders of calls and puts do not necessarily have to exercise them to earn a return.

They can sell them in the secondary market for their current value. The value of a call increases as the underlying common stock goes up in price. The value of a put decreases as the underlying common stock goes down in price.

Owners of call and put options can hedge by holding on to two or more securities to lower risk and at the same time make some profit. It may involve buying a stock and later purchasing an option on it. For example, a stock may be bought along with writing a call on it. Also, a holder of a stock that has risen in price may buy a put to furnish downside risk protection.

The writer of a call agrees to sell shares at the strike price for the price paid for the call option. Call option writers do the opposite of what buyers do. Investors write options because they believe that a price increase in the stock will be less than what the call purchaser expects. They may even expect the price of the stock to remain static or to decrease. Option writers receive the option premium minus related transaction costs. If the option is not exercised, the writer earns the price paid for it. However, when an option is exercised, the writer suffers a loss, sometimes a quite significant one.

When the writer of an option decides to sell shares, he must come up with stock at the agreed upon price if the option is exercised. In either case, the option writer receives income from the premium. (Shares are sold in denominations of 100.) An investor usually sells an option when he or she expects it not to be exercised. The risk of option writing is that the writer, if uncovered, must buy stock or, if covered, loses the gain.

A writer can buy back an option to terminate exposure. For example, assume the strike price is $40 and the premium for the call option is $5. If the stock is at less than $40, the call would not be exercised, and the investor must provide 100 shares at $40. However, the call writer would lose money only if the stock price exceeded $45.

Options may be naked (uncovered) or covered. Naked options are options on stocks that the writer does not own. The investor writes the call or put for the premium and will keep it if the price change is in his or her favor or is immaterial in amount. The writer's loss exposure is unlimited, however. Covered options are written against stocks the writer owns and are not quite as risky. For example, a call can be written for stock the writer owns or a put can be written for stock sold short. This is a conservative mechanism to obtain positive returns. The goal is to write an out-of-the-money option, keep the premium paid, and have the market price of the stock equal but not exceed the option exercise price. Writing a covered call option is similar to hedging a position because if the stock price falls, the writer's loss on the stock is partly netted against the option premium.

How Is It Used for Investment Decisions?   The major factors affecting the price of an option are the exercise price, the time premium, and the price of the underlying security.

An investor who feels that a stock may rise in value can buy a call option at a substantially lower price than the price of the stock with substantially higher leverage, risking only the cost of the call. Conversely, if the investor believes the stock will fall in price, a put option can be purchased for a fraction of the cost of the stock and profit when the stock declines in value with similar leverage.

Option strategies can be used for the purposes of increasing leverage, hedging risk, and improving the rate of return. This is accomplished by utilizing a call option writing strategy.

Call option writing strategies are also a very conservative method of investing because the premium from the call options provides a substantial hedge for a portfolio for downward price movements.

VALUATION OF STOCK WARRANTS

What Is This Tool?   A warrant is an option to purchase a certain number of securities at a stated price for a given time period at an exercise (subscription) price that is higher than the current market price. A warrant may or may not come in a one-to-one ratio with stock owned. Unlike an option, a warrant is usually good for several years; some, in fact, have no maturity date.

A warrant may be received as a sweetener when buying a bond or preferred stock.

How Is It Computed?   When warrants are issued, the exercise price is greater than the market price. Assume a warrant of R&J Co. stock enables an investor to buy one share at $25. If the stock exceeds $25 before the expiration date, the warrant increases in value. If the stock goes below $25, the warrant loses its value.

The exercise price for a warrant is usually constant over the warrant's life. However, the price of some warrants may rise as the expiration date approaches. Exercise price is adjusted for stock splits and large stock dividends. The return on a warrant for a holding period of no more than one year equals

$$\frac{\text{Selling price} - \text{acquisition price}}{\text{Acquisition price}}$$

Assume that an investor sells a warrant at $21. That same warrant cost him only $12. The return is

$$\frac{\$21 - \$12}{12} = \frac{\$9}{12} = 75 \text{ percent}$$

The return on a warrant for a holding period in excess of one year equals

$$\frac{(\text{Selling price} - \text{acquisition price})/\text{years}}{\text{Average investment}}$$

Assume that there is a holding period of four years on the warrant the investor just sold for $21. The return is

$$\frac{(\$21 - \$12)/4}{(\$21 + \$12)/2} = \frac{\$2.25}{\$16.50} = 13.6 \text{ percent}$$

The value of a warrant is greatest when the market price of the related stock is equal to or greater than the exercise price of the warrant. The value of a warrant thus equals (market price of common stock − exercise price of warrant) × number of common stock shares bought for one warrant.

For example, suppose that a warrant has an exercise price of $25. Two warrants equal one share. The market price of the stock is $30. The warrant has a value of

$$(\$30 - \$25) \times 0.5 = \$2.50$$

Usually the market value of a warrant is greater than its intrinsic value, or premium, because of the speculative nature of warrants. Premium equals the market price of the warrant minus its intrinsic value. For example, if the warrant referred to previously has a market price of $4, the premium is $1.50.

Example 1:   An investor holds eight warrants. Each warrant permits the investor to purchase one share of stock at $12 for one year. The warrant will have no value at the issue date if the stock is selling below $12. If the stock increases in value to $25 a share, the warrant should be worth about $13. The eight warrants will, thus, be worth approximately $104.

Example 2:   Assume MSL Co. common stock is $40 per share. One warrant can be used to buy one share at $34 in the next three years. The intrinsic (minimum) value per warrant is $6 − ($40 − $34) × 1. Because the warrant has three years left and can be used for speculation, it may be traded at an amount higher than $6. Assuming the warrant was selling at $8, it has a premium of $2. The premium is the $2 difference between warrant price and intrinsic value.

Even when the stock is selling for less than $34 a share, there might be a market value for the warrant because speculators may wish to buy it on the expectation of an attractive increase in common stock price in the future. For instance, if the common stock was at $30, the warrant has a negative intrinsic (minimum) value of $4, but the warrant might have a dollar value of, for example, $1 because of an expected rise in common stock value.

Example 3:   An investor may use the leveraging effect to boost dollar returns. Assume he or she has $7,000 to invest. If he or she buys common stock when the market price is $35 a share, this investor can buy 200 shares. If the price increases to $41 a share, the investor will have a capital gain of $1200, but if he or she invests the $7000 in warrants priced at only $7 a share, he or she can acquire 1000 of them. (One warrant equals one share.) If the price of the warrants increases by $6, profit will be $6000. In this instance, the investors earns a return of only 17.1 percent on the common stock investment whereas on the warrants the investor would get a return of 85.7 percent.

On the other hand, assume that the price of the stock drops by $6. If he or she invests in the common stock, he or she will lose $1200 for a remaining equity of $5800. However, if he or she invests in the warrant, he or she will lose everything (assuming no warrant premium exists).

Example 4:   Assume that an investor sells short 100 shares at $15 each and then buys warrants for 100 shares at $13 a share. The cost of the option is $3, or 3 points a share, a total of $300. In effect, he or she is buying the stock at $16 a share. Thus, if the stock rises above $15, the loss is limited to $1 a share.

Where Is It Found?   Prices of actively traded warrants may be found in stock tables in many major newspapers. Quotes on thinly traded warrants may be best found through electronic quotes services like telephone based ones provided by major brokerages to their clients or the quotes services in online computer databases. The exact terms of the warrants, however, often must be obtained from the company, brokers, financial publications such as *Value Line,* or through Securities and Exchange Commission filings on the Internet at www.sec.gov.

How Is It Used and Applied?   Warrants are not available for all securities. They pay no dividends and carry no voting privileges. The warrant enables the holder to take part indirectly in price appreciation of common stock and to obtain a capital gain. One warrant usually equals one share, but in some cases more than one warrant is needed to get one share.

Warrants can be bought from a broker or they are given to shareholders by a company as a fund-raising device. The price of a warrant is usually listed along with that of the common stock of the company.

Advantages of warrants are

- The price change in a warrant follows that of the related common stock, making capital gain possible.
- The low unit cost allows a leverage opportunity in the form of lowering the capital investment without damaging the investment's capital appreciation. This increases the potential return.
- Lower downside risk potential exists because of the lower unit price.

Disadvantages of warrants are

- If no price appreciation occurs before the expiration date, the warrant loses it value.
- The warrant holder receives no dividends.
- Investment in warrants requires sophistication.

How Is It Used for Investment Decisions?   When the price per common share goes up, the holder of the warrant may either sell it (because the warrant also increases in value) or exercise the warrant and get the stock. Trading in warrants is speculative. There is because potential for high return, but high risk exists because of the possibility of variability in return.

If an investor is to get maximum price potential from a warrant, the market price of the common stock must equal or exceed the warrant's exercise price. Also, lower-priced issues offer greater leverage opportunity. Furthermore, a warrant with a low unit price generates higher price volatility and less downside risk and, thus, is preferable to a warrant with a high unit price.

Warrants are speculative because their value depends on the price of the common stock for which they can be exchanged. If stock prices fluctuate widely, the value of warrants will sharply vacillate.

VALUE OF STOCK RIGHTS

What Is This Tool?   Some common stock owners have the preemptive right, which allows them to maintain their proportionate share in the company. Thus, they can buy new shares issued before they go on sale to the general public. This way they can maintain their percentage of ownership. One right is issued for each share of stock owned.

Rights typically have a life of no more than three months, with several weeks being customary.

How Is It Computed?   Because a stock right gives the holder the right to purchase a stock at a fixed price below the current market price by a certain period of time, it has its own market value. The value of a right depends on whether the stock is traded rights-on or rights-off. In a rights-on trade, the stock is traded with rights attached so the investor who purchases a share receives the attached stock right. In a rights-off or ex-rights trade, the stock and its rights are separate from each other and are traded in different markets.

The formula to determine the value of one right needed to buy a share of stock has to be adjusted for the rights-on and ex-rights conditions.

Rights-on condition

$$\text{Value of one right} = \frac{\text{Market value of stock, rights-on } - \text{ subscription price}}{\text{Number of rights required to purchase one share } + 1}$$

Ex-rights condition

$$\text{Value of one right} = \frac{\text{Market value of stock, ex-rights } - \text{ subscription price}}{\text{Number of rights required to purchase one share}}$$

Where Is It Found? The information needed to determine the value of a stock right and its attractiveness is found in the rights offering announcement distributed by the company to its stockholders. The rights offering includes the subscription price, number of rights to be received, expiration date, and other relevant terms. The current market price of the stock may be found in the stock pages of newspapers.

Example 1: An investor owns 3 percent of MNO Co. If the company issues 5000 additional shares, he or she may receive a stock rights offerings chance to buy 3 percent, or 150 shares, of the new issue. This right enables him to purchase new common stock at a subscription price (sometimes called an exercise price) for a short time. This subscription price, or exercise price, is lower than the current market price of the stock.

Example 2: Assume the current market price of stock is $30 a share. The new share has an exercise price of $26. An investor needs two rights to obtain one new share. One ex-right equals

$$\frac{\$30 - \$26}{2} = \frac{\$4}{2} = \$2$$

Provided the stock price holds at around $30 a share, the right has a value of $2.

How Is It Used for Investment Decisions? A rights offering is important to current stockholders because it confers on them the right of maintaining equivalent financial control over a company despite a new stock offering, as well as offering them the ability to protect against dilution of the value of the shares during a new stock offering. Further, the rights offering may allow the purchase of stock at a discount below the current market price per share. Additionally, brokerage commissions may be saved.

A secondary market exists for those stockholders who do not wish to exercise their right to purchase additional stock and wish, instead, to sell the stock right. Normally, the market price of a right exceeds its theoretical value. It is not unusual for investors to "bid up" the price of the stock in anticipation of future performance. Because rights do offer excellent leverage, investors can earn higher returns by purchasing rights rather than the stock. As the expiration date of the stock right is approached, its premium market value corresponds more closely to its theoretical value.

An interesting aspect of stock rights issues is that a small percentage of stockholders, normally about 1.5 percent, will not exercise or sell their rights. The net result is that these stockholders lose substantial sums of money by not exercising or selling their stock rights.

A Word of Caution: A stock right should not be exercised in a financially troubled company whose shares would be expected to decrease in price.

Tool #**81**

Profitability Tools

EARNINGS SURPRISES

What Is This Tool?   An "earnings surprise" occurs when a company reports profits or losses that differ from brokerage house analysts' expectations. A surprise can be either positive (higher than expected) or negative (lower than expected).

How Is It Computed?   Four companies, Zacks, First Call, I/B/E/S, and Nelson's vie for the title of being the leading trackers of analysis' earnings projections. These firms constantly poll brokerages for their earnings estimates. From that survey, these companies publish a compilation that includes the high, low, and mean prediction for a company's upcoming quarterly and fiscal year results.

When earnings are announced by the companies, these tracking companies then rate the "surprise" factor of the reported earnings as a percentage of the expected mean earnings. These companies also do other analyses including ranking companies by the percentage size of the surprises.

Where Is It Found?   Earnings expectations are available on many electronic quotation services such as Bloomberg Business News. *The Wall Street Journal* publishes a short list on notable surprises each day along with its daily listings of quarterly corporate results. Three other services also track such data, I/B/E/S, Nelson's, and First Call. Many brokerage houses have information from these two services available. On the Internet, Zacks Web site at www.zacks.com is an excellent free resource. (See Figure 77.)

How Is It Used for Investment Decisions?   From a macro viewpoint, earnings surprises can be used as a way to view the potential strength of the economy. A series of disappointing results from well-known companies could be a sign of overall weakness. Conversely, a string of profit news besting Wall Street's expectation is a signal that the national financial fortunes may have turned up.

On a microeconomic view, negative earnings surprises can have a devastating impact on stock prices, especially when they involve issues of fast-growing companies. Many institutional investors are known to dump shares soon after downside earnings surprises because they see it as a sign of potential trouble or because the traders may have lost confidence in their own abilities to project the company's future.

On the other hand, upward surprises can be beneficial for a company's stock because it reflects a positive earnings picture for the company.

```
┌─────────────────────────────────────────────────────┐
│ EPS Surprises                              ? _ X     │
│ Sponsored By:                                        │
│ SALOMONSMITH BARNEY                                  │
│                                                      │
│                    Positive Surprises                │
│  Ticker   Time    Expected    Reported      %        │
│                     EPS         EPS      Surprise    │
│   THC    09:14      0.45        0.48       6.67      │
│                                                      │
│                                                      │
│                   Negative Surprises                 │
│  Ticker   Time    Expected    Reported      %        │
│                     EPS         EPS      Surprise    │
│   SCHN   12:34      0.25        0.23        -8        │
│ Last updated on Oct 3, 2000 13:12:39                 │
└─────────────────────────────────────────────────────┘
```

FIGURE 77   How analysts' estimates are tracked by Zacks on the Internet. (From www. Zacks.com.)

HORIZONTAL (TREND) ANALYSIS

What Is This Tool?   Horizontal analysis is a time series analysis of financial statement items covering more than one year. It looks at the percentage change in an account over time.

How Is It Computed?   The percentage change equals the dollar change divided by the base year amount.

$$\text{Percentage change} = \frac{(\text{Year 2} - \text{Year 1})}{\text{Year 1}}$$

Example: If a company's sales in 19X1 and 19X2 were $5,000,000 and $4,000,000, respectively, there is a 20 percent decrease ($1,000,000/$5,000,000). The significant deteriorating sales position of the company appears to make it unattractive for investment.

Where Is It Found?   Most companies report horizontal percentage changes in their annual reports. Figure 78 shows the percentage changes in the financial highlights section of Procter & Gamble's 1992 Annual Report. If not, the investor finds a company's financial statement in its annual report. He or she then computes the percentage change for what are considered to be important items such as net income and sales.

The percentage change in market price of stock may be obtained by using the year-end market price per share, published in the stock quotations section in financial newspapers.

How Is It Used for Investment Decisions?   By examining the magnitude of the direction of a financial statement item, the investor can evaluate whether the company's financial position is getting better or worse. For example, if there is a significant growth trend in earnings, the investor may be advised to buy the stock.

A Word of Caution: Past trends in price may not necessarily predict the trend in future prices. The environment may have changed.

Also See: Profitability: Vertical (Common-Size) Analysis.

Restaurant Margin–Domestic

	1997	1996	1995
Company Sales	100.0%	100.0%	100.0%
Food and paper	31.3	32.1	32.3
Payroll and employee benefits	30.3	30.0	29.5
Occupancy and other operating expenses	27.0	27.4	27.1
Restaurant margin	11.6%	10.6%	11.2%

FIGURE 78   How one company breaks out profit margins in their annual report.

PROFITABILITY RATIOS

What Are These Tools?   Profitability ratios look at the company's earnings relative to sales and assets employed. They are important operating performance measures.

How Are They Computed?   Net profit margin is a ratio that reveals the profitability generated from sales. The higher the profit margin from each sales dollar generated, the better the company is doing financially.

$$\text{Net profit margin} = \frac{\text{Net income}}{\text{Net sales}}$$

The gross profit margin ratio is helpful in appraising a company's ability to effectively use its asset base to generate revenue.

$$\text{Gross profit margin} = \frac{\text{Gross profit}}{\text{Net sales}}$$

$$\text{Gross profit} = \text{Sales} - \text{Cost of sales}$$

$$\text{Net sales to total assets} = \frac{\text{Net sales}}{\text{Average total sales}}$$

Return on investment or ROI measures the effectiveness of the company's assets to create profits. Are assets being used productively?

$$\text{Return on investment (ROI)} = \frac{\text{Net income}}{\text{Average total assets}}$$

where

$$\text{Average total assets} = \frac{\text{Total assets (beginning)} + \text{Total assets (ending)}}{2}$$

ROI is the product of two important factors, net profit margin and total asset turnover.

$$\text{ROI} = \frac{\text{Net income}}{\text{Average total assets}} = \frac{\text{Net income}}{\text{Sales}} \times \frac{\text{Sales}}{\text{Average total assets}}$$

$$= \text{Net profit margin} \times \text{total asset turnover}$$

The return on stockholders' equity or ROE ratios reveals the earnings earned by stockholders in the business.

$$\text{Return on stockholders equity (ROE)} = \frac{\text{Net income}}{\text{Average stockholders equity}}$$

Where Are They Found?   The profitability and rate of return ratios appear in brokerage research reports, such as *Value Line Investment Survey,* prepared by fundamental analysis. They are sometimes mentioned in management's discussion within the annual report. In any event, the investor may easily calculate these ratios from the financial information contained in the balance sheet and income statement published in the company's annual report. Online computer investment database such as www.marketguide.com contains such information.

Example 1:   A company reports the following information:

	20X2	20X1
Gross profit	$15,000	$20,000
Net income	8,000	9,600
Sales	65,000	80,000
Relevant ratios follow.		
Net profit margin	0.12	0.12
Gross profit margin	0.23	0.25

The net profit margin was constant, indicating that the earning power of the business remained static. However, there was an improvement in gross profit, probably owing to increased sales and/or control of costs of sales. The reason the gross profit margin is up but net profit margin is constant, even though sales increased, is probably owing to a lack of control in operating expenses.

Example 2:

	20X2	20X1
Net income	$259,358	$384,346
Average total assets		
Beginning of year	1,548,234	1,575,982
End of year	1,575,982	1,614,932
Average total assets	1,562,108	1,595,457
Return on total assets	16.50%	24.09%

There has been growth in the return on assets over the year, indicating the assets' greater productivity in generating earnings.

How Are They Used for Investment Decisions?　An indication of good financial health is a company's ability to earn a satisfactory profit and return on investment. The investor should be reluctant to associate himself with an entity that has poor earnings potential because the market price of stock and future dividends will be adversely affected.

By examining a company's profit margin relative to previous years and to industry norms, the investor can evaluate the company's operating efficiency and pricing strategy as well as its competitive status within the industry. The profit margin may vary greatly within an industry because it is subject to sales, cost controls, and pricing. (See Figure 79.)

ROI and ROE are used to measure a company's success and to rank companies in the same industry.

A Word of Caution:　Profitability ratios may appear attractive but they may be overstated because the company has manipulated its earnings, reduced discretionary costs needed for future growth, or obtain one-time earnings boosts.

Return on Equity for Stocks

Company Name	Ticker	Exchange	ROE 1997–1998	ROE 1996–1997
Dell Computer	DELL	NASDAQ	73	64.3
Coca-Cola	KO	NYSE	56.5	56.7
Computer Associates International	CA	NYSE	47.1	24.4
Bristol-Myers Squibb	BMY	NYSE	44.4	43.4
Philip Morris Companies	MO	NYSE	42.3	44.3
Abbott Laboratories	ABT	NYSE	41.9	39
Global Marine	GLM	NYSE	38.6	39.2
Merck	MRK	NYSE	36.6	32.4
Intel	INTC	NASDAQ	36	30.6
General Motors	GM	NYSE	35.9	19.6
Parametric Tech	PMTC	NASDAQ	34	26.9
CIENA	CIEN	NASDAQ	31.1	32.4
PepsiCo	PEP	NYSE	30.9	17.3
Warner-Lambert	WLA	NYSE	30.7	30.5
IBM	IBM	NYSE	30.6	25
Texas Instruments	TXN	NYSE	30.5	1.5
Amgen	AMGN	NASDAQ	30.1	35.7
Procter & Gamble	PG	NYSE	30	27.5
Gillette	G	NYSE	29.5	21.2
Tellabs	TLAB	NASDAQ	28.3	20

FIGURE 79　Here is a look at the highest return on equity figures for fiscal years ending in 1998 by October 31, 1998 for actively traded (more than 2 million shares daily average) U.S. stocks. (From Morningstar Inc. With permission.)

Also See: Profitability: Quality of Earnings, Share–Price Ratios: Cash per Share, Share–Price Ratios: Earnings per Share.

QUALITY OF EARNINGS

What Is This Tool? Quality of earnings are the realistic earnings of a company that conform to economic reality. Quality of earnings is a multifaceted concept that embraces many accounting and financial considerations and involves qualitative and quantitative elements. Qualitative elements such as cash flow are subject to measurement. Quantitative elements such as the quality of management cannot be measured objectively. This discussion will look only at the quantitative aspect because that is subject to computation. Quality of earnings can be analyzed only by a sophisticated investor.

How Is It Computed?

Reported net income + Items unrealistically deducted from earnings −
Items unrealistically added to earnings
    = Quality of earnings

There is no absolute "true" (real) earnings figure. However, the "quality of earnings" figure (adjusted earnings) should be more representative of the company's operational activity than reported net income.

$$\text{Salomon Brothers' Earnings Quality Index} = \frac{\text{Economic profits}}{\text{Net income}}$$

Salomon Brothers defines economic profits as reported profits adjusted to remove inventory profits and inadequate depreciation. A high ratio indicates better quality of earnings.

Example 1: A company reports sales of $1,000,000 and net income of $400,000. Included in the net income figure is research and development expense of $50,000, or 5 percent of sales. However, in past years the company's research and development relative to sales was 8 percent. Competing companies are showing 8 percent this year as well. Thus, the investor can conclude that research and development should be realistically $80,000 (8 percent × $1,000,000). Hence, R&D is understated by $30,000 ($80,000 - $50,000). The adjusted earnings follows.

Reported net income	$400,000
Less the understatement of R&D	−30,000
Quality of earnings	$370,000

In this example, there was only one adjustment. Of course, many adjustments would typically be required.

Example 2: Assume a company's earnings per share (EPS) of $6.00 includes some low quality components. These items are listed below as deductions from reported EPS. In order to arrive at an "acceptable" quality EPS, certain items must be deducted.

These were chosen with a view toward developing an approach that allows for a clearer understanding of the adjustment process. In reality, of course, reported EPS would be adjusted upward or downward for various reconciling items. An example of an upward adjustment would be the adding back to EPS of the effect of an unjustified accounting cushion arising from overestimated warranty provisions or bad debt provisions.

Reported EPS	$6.00
Deductions from reported EPS in order to arrive at an "acceptable quality" EPS	
Unjustified cutbacks in discretionary costs (e.g., advertising, repairs) as a percentage of sales	0.03
Extraordinary gains (e.g., sales of real estate)	0.04
A decline in the warranty provision that is not consistent with previous experience	0.02
Increase in deferred expenditures that do not have future economic benefit	0.01
"Acceptable quality" EPS	$5.90

Where Is It Found?   Some investment advisory research reports are solely devoted to the analysis and computation of a company's quality of earnings.

The knowledgeable, sophisticated investor also may determine a company's quality of earnings by analyzing financial information released by the company in its annual report and SEC filings (e.g., Form 10-K). There are books devoted to evaluating and determining a company's quality of earnings that the investor may refer to. Examples are *Quality of Earnings: The Investor's Guide to How Much Money A Company Is Really Making,* by Thorton O'Glove (Macmillan, New York, 1987), *How to Analyze Businesses, Financial Statements, and the Quality of Earnings,* by Joel Siegel (Prentice-Hall, New Jersey, 1991, 2nd ed.), and *Financial Statement Analysis,* by Leopold Bernstein (Richard D. Irwin, Illinois, 1989, 4th ed.).

How Is It Used and Applied?   Quality of earnings is the result of many factors including accounting policies used, adequacy of repairs and maintenance, stability in operations and earnings, accounting changes, income manipulation, appropriateness of deferring costs, underaccrued or overaccrued expenses, revenue recognition methods, adequacy of discretionary costs, degree of accounting estimates, inflationary profits, business risk, cash earnings, residual income (net income less minimum return on total assets), sales returns, and extent of diversification.

How Is It Used for Investment Decisions?   Earnings quality is relative rather than absolute; it applies to comparing the characteristics of net income among companies in the same industry. The investor should note the following:

- The "quality of earnings" encompasses much more than the mere understatement or overstatement of net income; it also refers to such factors as the stability of income statement components, the realization risk of assets, and the maintenance of capital.
- Quality of earnings affects the market price of stocks and bonds, dividends, and credit rating.

- Identical earnings of competing companies may possess different degrees of quality. The key to evaluating a company's earnings quality is to compare its earnings profile (the mixture and the degree of favorable and unfavorable characteristics associated with reported results) with the earnings profile of other companies in the same industry. Investors attempt to assess earnings quality in order to render the earnings comparable and to ascertain the value to be placed on such profits.

"Poor earnings quality" occurs when the company's accounting policies are not realistic.

A Word of Caution: The analysis of the quality of earnings should be performed only by investment professionals.

Also See: Profitability: Profitability Ratios, Share–Price Ratios: Cash per Share, Share–Price Ratios: Earnings per Share.

VERTICAL (COMMON-SIZE) ANALYSIS

What Is This Tool? In vertical analysis, a financial statement item is used as a base value. All other accounts in the financial statements are compared to it.

How Is It Computed? In the balance sheet, total assets equal 100 percent. Each asset is stated as a percentage of total assets. Similarly, total liabilities and stockholders' equity are assigned 100 percent with a given liability or equity account stated as a percentage of the total liabilities and stockholders' equity.

In the income statement, 100 percent is assigned to net sales with all revenue and expense accounts related to it.

Where Is It Found? Most companies report common-size percentages in their annual reports. If not, the investor finds a company's financial statement figures in its annual report, and then computes the vertical percentages.

Example 1:

Net sales	$300,000	100%
Less the cost of sales	60,000	20%
Gross profit	$240,000	80%
Less the operating expenses	150,000	50%
Net income	$90,000	30%

Example 2:

Current assets	$200,000	25%
Noncurrent assets	600,000	75%
Total assets	$800,000	100%

How Is It Used for Investment Decisions? Common size analysis can be compared from one period to another to see how the company is doing.

Vertical analysis tends to exhibit the internal structure of the enterprise. It indicates the relative amount of each income statement account to revenue. It shows the mix of assets that produces the income and the mix of the sources of capital, whether provided by current or long-term liabilities, or by equity funding.

The vertical percentages of a company should be compared to its competitors or to industry percentages so that the investor may ascertain the firm's relative position.

If vertical analysis indicates improved financial condition, such as increasing profit relative to sales, the investor may consider buying the stock.

A Word of Caution: The percentage relationship of an item to sales in one year may drastically change in another year. For example, a company's profit margin (net income to sales) may go from 30 percent in one year to 2 percent in the next year. One cause might be a recession.

Also See: Profitability: Horizontal (Trend) Analysis.

TOOL #82

REAL ESTATE TOOLS

CAPITALIZATION RATE

What Is This Tool? The capitalization rate—also known as "cap rate" or "income yield"—is a method used to determine the rate of return on a real estate investment.

How Is It Computed? The capitalization rate equals net operating income (NOI) divided by the purchase price. Assume NOI is $25,000 and the investment was $200,000. The capitalization rate equals

$$\frac{\$25,000}{\$200,000} = 12.5 \text{ percent}$$

If the market rate is 10 percent, the fair market value of similar property is $250,000 ($25,000/10 percent). The property may be underpriced.

Where Is It Found? The capitalization rate may be found on the real estate broker's fact sheet on a property.

How Is It Used for Investment Decision? The capitalization rate when applied to the earnings of an investment determines its market value. The lower the capitalization rate, the higher the anticipated risk to the investor and the higher the asking price paid. In determining whether a piece of property is underpriced or overpriced, the investor should look at the capitalization rate of similar kinds of property in the marketplace. The investor must note two limitations with this appraisal approach. First, it is based on only the first year's NOI. Second, the method ignores return through appreciation in property value.

Also See: Gross Income Multiplier (GIM), Real Estate Returns: Net Income Multiplier (NIM).

HOME PRICE STATISTICS

What Are These Tools? There are several measures reported monthly.

1. New home activity, a monthly count of sales of newly constructed homes.
2. Home resale activity, a monthly check on sales of existing housing.
3. Home affordability, a quarterly measure of prices vs. income and interest rates.

How Is It Computed? New home sales, both average price and sales volume, are reported by the U.S. Commerce Department. Sales figures are adjusted to reflect the seasonality of activity. Both sales and price figures are reported on both a national and regional basis.

Home resale activity, both average price and sales volume, are reported by the National Association of Realtors (NAR). Sales figures, too, are adjusted for seasonality. Both sales and price figures are reported on a national and state basis as well as for many major metropolitan areas.

The NAR's affordability index shows how likely it is for an average American family to afford to buy an existing home, considering current home prices, income levels, and mortgages rates. NAR's affordability index was 132.5 during the third quarter of 1998, meaning that one-half of the nation's households had at least 132.5 percent of the income needed to purchase a home at the third quarter median price, which was $132,700. Under these conditions, a family earning the median income of $45,281 could afford a home costing $175,800.

Where Is It Found? Such housing figures appear regularly in newspapers such as *The Wall Street Journal, The New York Times,* and *USA Today.* They are reported on business news TV and can be found on the Internet at the government statistics web site at www.stats.bls.gov or at NAR's www.realtor.com. (See Figure 80.)

How Is It Used for Investment Decisions? These figures can be used to judge the health of both the overall economy as well as the housing industry.

New home sales figures, particularly volume, are seen as a key barometer of future economic growth, because the housing and construction sector often leads the nation out of an economic slump. They can also be viewed as a way to determine the potential for home builders' shares as an investment.

Because of an increase in entry-level home buying, the rise in media understates the appreciation rate, which is now around 10 percent, D

(chart)

All Homes	No Sold Aug-99	No Sold Aug-00	Pct. Chng.	Median Aug-99	Median Aug-00	Pct. Chng.
Los Angeles	10,250	10,443	1.9%	$191K	$205K	7.3%
Orange County	4,612	4,729	2.5%	$241K	$274K	13.7%
San Diego	4,941	4,798	-2.9%	$213K	$237K	11.3%
Riverside	3,461	3,918	13.2%	$150K	$162K	8.0%
San Bernardino	3,031	3,471	14.5%	$137K	$147K	7.3%
Ventura	1,492	1,380	-7.5%	$238K	$251K	5.5%
So. California	27,787	28,739	3.4%	$195K	$211K	8.2%

Source: DataQuick Information Systems

FIGURE 80 How a web site tracks existing home sales activity. (From DataQuick.com.)

Existing home sales figures, notably price changes, are seen more as an indicator of consumer sentiment, another key factor because individuals purchase two-thirds of the nation's goods and services. Rising volumes, as well as strong price showing, make homeowners feel better about their own economic outlook and motivates them to spend.

The affordability index is a way to judge current and future potential for the housing industry. Low affordability can foreshadow upcoming problems for housing as more potential buyers stay out of the market.

A Word of Caution: Housing figures can be fairly erratic and are highly regionalized. That makes them difficult to extrapolate to major national trends. On a local level, however, the small sample base—only a tiny portion of homes in any market change hands in a year—makes sales prices figures difficult to use for an individual home.

Also See: Economic Indicators: Housing Starts and Construction Spending.

NET INCOME MULTIPLIER (NIM)

What Is This Tool? The net income multiplier (NIM) is a method to determine the price of income-producing property.

How Is It Computed? The multiplier equals the asking price (or market value) of the commercial property divided by the current net operating income (NOI). NOI equals the gross rental income less allowances for vacancies and operating expenses, except for depreciation and interest on debt. Assume that net operating income is $20,000 and the asking price is $200,000. The NIM equals

$$\frac{\$200,000}{\$20,000} = 10$$

If similar commercial property in the locality is selling for "8 times annual net," the value would be taken as $160,000 (8 × $20,000). This means that the property is overvalued and should *not* be bought.

Where Is It Found? The NIM for commercial property in an area may be ascertained by inquiring with real estate agents and reading published real estate information. The real estate broker's fact sheet on a property typically provides the NIM. The investor also should obtain an understanding of the real estate market in the locality by asking around and finding out what similar property has been selling for in the market.

How Is It Used for Investment Decisions? The NIM method is used by the investor to determine the approximate market value of a property. It is superior to the gross income multiplier (GIM) approach because it considers vacancies and operating expenses. A property may be purchased when it is undervalued. However, if the investor currently owns the property, he should sell it if it is overvalued before the property declines in price. For example, if the property's NIM is "8 times annual net" but the going market rate is 5 times annual net, it is overpriced and should be sold.

Also See: Real Estate Returns: Capitalization Rate (Cap Rate, Income Yield), Gross Income Multiplier (GIM).

NCREIF PROPERTY PERFORMANCE AVERAGES

What Is This Tool?   These averages are market-value property performance averages broken down by building type and geographic region.

How Is It Computed?   The index, compiled by the National Council of Real Estate Investment Fiduciaries, measures investment results over time for nonleveraged investment grade warehouse-research-and-development-office facilities, retail properties, office buildings, and apartment buildings. Properties in the index are all held in pension investment pools. There are subindexes for each. There are also two distinct parts to the index and its subindexes: one measures income, the other appreciation. In addition, national and eight subregional indexes are compiled.

A determination is made through appraisals of the average market values of property in selected major geographical areas. The average income generated from the properties is also determined by the company.

The index started on December 31, 1977. The apartment return index began in 1988.

Where Is It Found?   It appears in *Barron's*. (See Figure 81.)

How Is It Used for Investment Decisions?   The investor can use this information to decide lucrative areas in the country for real estate investment. Further, the data may be used to identify negative trends prompting a sell decision now before property values deteriorate further. The index can also be used to view broader trends in real estate prices, notably for those considering a real estate investment trust (REIT).

A Word of Caution: The appraisal of what a property is worth has subjectivity associated with it. Further, property values and income thereon may drastically change depending on economic and demographic conditions.

REAL ESTATE PERFORMANCE AVERAGES

	2000 Mar.	1999 Dec.	1999 Sep.	1999 Jun.	1999 Mar.
NCREIF Prop Index(Mil$)	82,820.8	77,432.3	75,676.6	72,780.8	68,056.8
Apts.	17.1%	16.9%	16.3%	16.4%	16.6%
Industrial	16.1	16.6	16.3	16.0	16.1
Office	42.0	41.9	42.6	42.2	40.7
Retail	23.5	24.3	23.5	24.1	25.4
East	28.5	28.7	28.5	28.6	28.8
Midwest	16.1	16.0	16.6	16.8	16.3
South	22.2	22.1	21.8	22.1	21.6
West	33.2	33.2	33.0	32.6	33.3
Total Return	2.31	2.94	2.77	2.60	3.59
Wilshire RE Index(Mil$)	113,460	113,090	115,643	130,676	116,974
Total Return	2.81%	0.23%	(9.54)%	10.62%	(3.47)%
Dividend Return	1.79	2.28	1.54	1.58	1.66
Wilshire REIT Index					
Total Return	2.99	0.17	(8.27)	10.56	(4.10)
Dividend Return	1.91	2.49	2.11	1.73	1.76
Wilshire REOC Index					
Total Return	0.88	0.84	(19.95)	10.97	3.44
Dividend Return	0.69	0.30	0.49	0.37	0.55

Sources: National Council of Real Estate Investment Fiduciaries, Two Prudential Plaza, 180 N. Stetson Avenue, Suite 2515, Chicago, IL 60601. Wilshire Asset Management, 1299 Ocean Avenue, Suite 700, Santa Monica, CA 90401–1085.

FIGURE 81   Here's how newspapers cover real estate performance data. (From *Barron's Market Week,* August 28, 2000, p. MW78. With permission.)

Tool #**83**

Russell 2000. Index

What Is This Tool? The Russell 2000 Index is seen as the top barometer of performance for small company stock in the U.S. It has received growing attention in the late 1990s as large stock consistently outperformed smaller ones.

How Is It Computed? The index is calculated by Frank Russell Co. of Seattle. Russell first ranks all U.S. stocks from the largest to smallest based on market capitalization each May 31. Russell excludes stocks trading below $1.00; it is, generally limited to only one class of a company's stock and does not include closed-end mutual funds, limited partnerships, royalty trusts, non-U.S. domiciled stocks, foreign stocks, and the like.

The 3000 largest stocks become the Russell 3000 Index, which tracks the broad U.S. market. The largest 1000 of those 3000 become the Russell 1000 Index, which tracks large U.S. stocks. The remaining 2000 is the roster of the Russell 2000.

Where Is It Found? Russell indexes appear in major daily newspapers as well as in *The Wall Street Journal* and *Barron's.* You can also track the index on major online computerized investment services or at Frank Russell Co.'s Web site at www.russell.com.

How Is It Used for Investment Decisions? Russell indexes can be used to detect the quality of the second tier market. Many studies have shown that over time small company stocks have performed better than large company shares.

The Russell 2000 represents approximately 10 percent of the total market capitalization of all U.S. shares. As of May 31, 1998, the median market capitalization was approximately $500.0 million and the largest company in the index had an approximate market capitalization of $1.4 billion.

An investor looking to catch the next upswing in small company stock might watch the relative performance of the Russell 2000 vs. the Russell 1000, the Dow Jones Industrial Average, or the S&P 500. When the Russell 2000 was consistently underperforming, that could be a buy signal.

The Russell 3000 is seen as a vane for the broad U.S. market. It also offers some interesting subindexes that track various industries.

- *Russell 3000 Autos and Transportation Index* comprises what are traditionally known as transportation companies plus auto companies.
- *Russell 3000 Consumer Discretionary and Services Index* comprises makers of products and providers of services directly to the customer.
- *Russell 3000 Consumer Staples Index* comprises companies that provide products directly to the consumer that are typically considered nondiscretionary items based on consumer purchasing habits.

- *Russell 3000 Financial Services Index* comprises financial-service firms.
- *Russell 3000 Health Care Index* comprises firms in medical services or health care.
- *Russell 3000 Integrated Oils Index* comprises oil companies in exploration, production, and refining processes.
- *Russell 3000 Materials and Processing Index* comprises companies that extract or process raw materials.
- *Russell 3000 Other Energy Index* comprises energy-related businesses other than integrated oils.
- *Russell 3000 Producer Durables Index* of companies that convert unfinished goods into finished durables used to manufacture other goods or provide services.
- *Russell 3000 Technology Index* comprises the electronics and computer industries or makers of products based on the latest applied science.
- *Russell 3000 Utilities Index* comprises utilities companies in industries heavily regulated.

A Word of Caution: Just because history tells you that small stocks like those that comprise the Russell 2000 have outperformed in the past, that is no guarantee that such a trend will continue forward. With the growth of money management in the 1990s, there are far less small, undiscovered stocks to invest in. Plus, with so many large money managers running huge mutual funds, there are less investors for small stocks. Most large funds cannot, or will not, purchase small stocks because they do not have enough liquidity to be easy to trade in and out of. (See Figure 82.)

Also See: Dow Jones Industrial Average, NASDAQ Indexes; Standard & Poor's (S&P) 500, Wilshire 5000 Equity Index.

Indexes

Index Tracks?	Dow Jones Industrial Average Large Cap	Standard & Poor's 500 Broad Market	NASDAQ Composite NASDAQ	Russell 3000 Broad Market	Russell 2000 Small Cap	Russell 1000 Large Cap
Companies in index?	30	500	5,500	3,000	2,000	1,000
How constructed/ weighted?	By price	By market capitalization	By market capitalization	By market capitalization	By market capitalization	By market capitalization
Criteria for a stock's inclusion?	Selection committee; to be representative of U.S. industry	Selection committee; to include biggest U.S. stocks	All domestic shares traded on NASDAQ's National Market and Small Cap Exchange	Top 3,000 NYSE, NASDAQ & AMEX domestic stocks, ranked by market capitalization	Smallest 2,000 stocks, by market cap, in Russell 3000	Largest 1,000 stocks, by market cap, in Russell 3000
Median market cap of index stocks?	$4.9 billion	$5 billion	Not available	$790 million	$500 million	$3.7 billion
Percent of index on NYSE, by market cap?	100%	89.5%	0	83.9%	50.6%	87.2%
Percent of index on NASDAQ, by market cap?	0	10.2%	100%	15.5%	46.4%	12.4%
Percent of index on AMEX, by market cap?	0	0.3%	0	0.7%	3%	0.4%
Maximum time between changes in index members?	None	None	None	1 year	1 year	1 year
10-year performance through 10/31/98?	17.26%	17.28%	15.89%	16.46%	11.15%	16.99%

FIGURE 82   Russell Indexes vs. Other Key Indexes, as of November 1998.

TOOL #**84**

SAFETY AND TIMELINESS RANKING

What Is This Tool?   Investment advisory services monitor and rate hundreds of securities in terms of safety and timeliness. Value Line has a good track record.

How Is It Computed?   Value Line uses a financial model directed toward determining a company's profit growth and estimates what earnings will be over the next year. The computerized model then projects which stocks will perform the best or worst in terms of price over the next 12 months. A risk rating is assigned to each security based on its historical fluctuation in price compared to a market index. This is measured by beta. Industries also are ranked.

Where Is It Found?   The rankings of stocks and industries may be found in *Value Line's Investment Survey* publication.

How Is It Used and Applied?   Value Line assigns one of the following rankings based on the timeliness and safety of the company's stock.

1. Best.
2. Above average.
3. Average.
4. Below average.
5. Worst.

How Is It Used for Investment Decisions?   The investor may use the Value Line information as a basis to buy or sell a stock. A company rated number 1 may be an attractive buy. However, a number 5 category stock should be avoided or sold if currently held.

Besides ranking companies and industries, Value Line includes other information including corporate financial data, institutional percentage of ownership of the company, and insider transactions.

A Word of Caution: The safety and timeliness ranking may be used along with beta, the price–earnings ratio, and other data such as the institutional percentage of ownership of the company and insider transactions.

Also See: Beta for a Security, *Investor's Business Daily*'s Intelligent Tables.

Tool #85

Share-Price Ratios

BOOK VALUE PER SHARE

What Is This Tool?   Book value per share is the net assets available to common stockholders divided by the shares outstanding, where net assets is stockholders' equity less preferred stock. It is what each share is worth based on historical cost.

How Is It Computed?   Book value per share of preferred stock equals

$$\frac{\text{Liquidation value of preferred stock } + \text{ preferred dividends in arrears}}{\text{Preferred shares outstanding}}$$

Book value per share of common stock equals

$$\frac{A - (B + C)}{D}$$

where

A equals total stockholders' equity.
B equals liquidation value of preferred stock.
C equals preferred dividends in arrears.
D equals common shares outstanding.

Care must be taken in computing the liquidation value of preferred stock. Some companies have preferred stock issues outstanding that give the right to significant liquidation premiums that substantially exceed the par value of such shares. The effect of such liquidation premiums on the book value of common stock can be quite material.

Example: The following information is given.

Total stockholders' equity $4,000,000.
Preferred stock, 6 percent dividend rate, 100,000 shares, $10 par value, $12 liquidation value.
Common stock, 200,000 shares, $20 par value.
Preferred dividends in arrears for 3 years.
Liquidation value of preferred stock equals 100,000 shares times $12 equals $1,200,000.
Preferred dividends in arrears equals par value of preferred stock 100,000 times $10 equals $1,000,000.

Preferred dividend rate times 6 percent.
Preferred dividend per·year equals $60,000
Number of years (times 3)
Preferred dividend in arrears equals $180,000.

Book value per share for preferred stock equals

$$\frac{\$1,200,000 + \$180,000}{100,000 \text{ shares}} = \frac{\$1,380,000}{100,000} = \$13.10$$

Book value per share for common stock equals

$$\frac{\$4,000,000 - \$1,380,000}{200,000 \text{ shares}} = \frac{\$2,620,000}{200,000} = \$13.10$$

Where Is It Found?   Book value per share is usually reported in financial advisory service publications (e.g., *Value Line Investment Survey, Standard & Poor's Stock Guide*) and brokerage research reports. Computer investment online services like www.marketguide.com and yahoo.com also contain such figures. It also may be calculated by the investor because all needed information is contained in the annual report.

How Is It Used for Investment Decisions?   A comparison of book value per share with market price per share gives an indication of how the stock market views the company. If market price per share is significantly below book value per share, the investment community is not favorably disposed toward the company's stock. However, the stock may be undervalued if the investor believes the company has future potential. Thus, it may be a buying opportunity.

A Word of Caution: Book value per share should be compared to peer firms. A stock trading at a discount to its book value may not automatically be a buying opportunity if shares of its competitors trade at discounts equal, or even greater, in scale. This may be an indication that traders expect this industry to have rough times ahead that will push book values lower in the future.

Also See: Share–Price Ratios: Earnings per Share, Share–Price Ratios: Price-Book Value Ratio.

CASH PER SHARE

What Is This Tool?   Cash per share is the per share cash earnings of the company. Earnings are of higher quality if they are backed by cash. Cash may be used to pay debt, buy back stocks or bonds, buy capital assets, pay dividends, and so on.

How Is It Computed?

$$\text{Cash per share} = \frac{\text{Cash flow from operations}}{\text{Total shares outstanding}}$$

Cash flow from operations equals net income plus noncash expenses (e.g., depreciation) minus noncash revenue (e.g., amortization of deferred revenue equals the cash flow from operations (cash earnings)).

Cash flow from operations may be approximated by the investor by adding back to net income depreciation.

Compute the ratio of cash earnings to net income. The trend in this ratio should be thoroughly examined. A higher ratio is desirable because it means that the net income is supported by the internal generation of cash. This is a cost-free source of financing.

Where Is It Found?   Cash earnings per share may be found in some brokerage research reports and financial advisory services (e.g., the *Value Line Investment Survey*). Computer investment online services like www.marketguide.com and yahoo.com also contain such figures. The investor also can compute it from information readily available in the company's annual report. Depreciation is simply added to net income and then divided by the shares outstanding at year-end.

Example: A company's net income for 19X1 is $5,700,000, which includes a depreciation expense of $300,000. There are 1,000,000 shares outstanding. The cash earnings per share equals

$$\frac{\$5,700,000 + \$300,000}{1,000,000} = \$6 \text{ per share}$$

How Is It Used for Investment Decisions?   Net income backed by cash is important because it represents a liquid source of funds. The investor should place a premium on a company's earnings that are supported by cash. Such earnings are worth more and should be reflected in a higher market price per share. Also, in cases when shares on a particular industry, for example, are extremely depressed, some stocks may trade at a discount to their cash-per-share ratio. That means the company is valued by traders at less than its cash holdings. In that case, the company may be a takeover candidate because a raider could, in theory, buy the company for free. That is, the target's cash would pay for the deal, giving the acquirer the remaining business essentially for nothing.

A Word of Caution: Cash per share may be artificially high when there are few shares outstanding even though the company's cash position is weak. Further, cash per share does not necessarily reflect a company's profitability as depicted by the earnings per share.

Also See: Share–Price Ratios: Earnings per Share.

EARNINGS PER SHARE

What Is This Tool?   Earnings per share (EPS) is the amount of the company's earnings to each share held by the investing public.

How Is It Computed?   Dual presentation of EPS is made as follows.

$$\text{Primary EPS} = \frac{A - B}{C + D}$$

$$\text{Fully diluted EPS} = \frac{A - B}{C + D + E}$$

where

A equals net income.
B equals preferred dividends.
C equals weighted-common stock outstanding.
D equals common stock equivalents.
E equals other fully diluted securities.

Weighted-average common stock outstanding takes into account the number of months in which those shares were outstanding.

To get fully diluted EPS, one needs to add to shares outstanding. Common stock equivalents are securities that can become common stock at a later date including stock options, stock warrants, and convertible securities (when the yield is less than 2/3 of the average AA-rated corporate bond yield at the time of issuance).

Other fully diluted securities are convertible securities with a yield equal to or greater than 2/3 of the average AA corporate bond yield at the time of issuance. In 1998, fully diluted EPS became the standard way that companies report their earnings.

Where Is It Found?　　A company's earnings per share are routinely included in news about corporate profits in daily newspapers, on business news TV broadcasts and in coverage through online computer investment services. It is also tracked in its quarterly and annual reports. In addition, brokerage reports and financial advisory services (e.g., the *Value Line Investment Survey*) report the company's earnings per share.

How Is It Used for Investment Decisions?　　The investor is interested in earnings per share as a measure of the profitability of the company that can be easily compared to previous reporting periods. A company with growing earnings per share has been successful in its operating performance. A higher earnings per share will likely result in higher dividends per share for those income-oriented firms and higher market price per share for all stocks. On the other hand, a declining or negative earnings per share infers financial problems negatively impacting the attractiveness of the company's stock.

The trend (momentum) in earnings per share should be examined as an indication of the company's earning power (see Share–Price Ratios: Growth Rate).

A Word of Caution:　　The earnings per share statistic should be used in conjunction with other ratios such as cash per share and dividend payout. Also, companies can use stock buybacks to inflate EPS without improving the actual profit level.

Also See:　　Share–Price Ratios: Cash per Share.

GROWTH RATE

What Is This Tool?　　The growth rate of a business may be expressed in terms of earnings, dividends, sales, market price, and assets. A higher premium is assigned to a company that has a track record of growth.

How Is It Computed?   Growth rate in earnings per share equals

$$\frac{\text{Earnings per share (end of period)} - \text{earnings per share (beginning of period)}}{\text{Earnings per share (beginning of period)}}$$

Assume that earnings per share for 20X1 and 20X2 were $1.25 and $1.50 per share, respectively. The annual growth rate in earnings per share equals

$$\frac{\$1.50 - \$1.25}{\$1.25} = 20 \text{ percent}$$

A 20 percent growth rate in earnings from 20X1 and 20X2 is favorable.

The same approach may be used in computing the growth rate in dividends per share. Other measures of growth also may be used, such as the change in sales.

Growth rate may be expressed in terms of a compounded annual rate equal to

$$\text{Compounded annual rate of growth} = F_n = P \times \text{FVIF}(i, n)$$

where

F_n equals the future value amount.
P equals the present value amount.
FVIF(i, n) equals the future value factor based on the interest rate (i) and number of periods (n).

Solving this for FVIF, the investor would obtain FVIF(i, n).

$$\text{FVIF}(i, n) = \frac{F_n}{P}$$

Assume that a company has earnings per share of $2.50 in 19X9 and ten years later the earnings per share has increased to $3.70. The compound annual rate of growth in earnings per share equals

$$F_{10} = \$3.70 \quad \text{and} \quad P = \$2.50$$

Therefore,

$$\text{FVIF}(i, 10) = \frac{\$3.70}{\$2.50} = 1.48$$

From a future value of $1 table, a FVIF of 1.48 at 10 years is at $i = 4$ percent. The compound annual rate of growth is, therefore, 4 percent. (See Figure 83.)

Where Is It Found?   The investor may find earnings per share, dividends per share, and sales in the company's annual report. From these published figures, the investor may compute the appropriate growth rates. In addition, brokerage research reports and financial advisory publications (e.g., the *Value Line Investment Survey*)

Future Value of $1.00

	1%	2%	3%	4%	5%	6%	7%	8%	9%	10%
1	1.010	1.020	1.030	1.040	1.050	1.060	1.070	1.080	1.090	1.100
2	1.020	1.040	1.061	1.082	1.103	1.124	1.145	1.166	1.188	1.210
3	1.030	1.061	1.093	1.125	1.158	1.191	1.225	1.260	1.295	1.331
4	1.041	1.082	1.126	1.170	1.216	1.262	1.311	1.360	1.412	1.464
5	1.051	1.104	1.159	1.217	1.276	1.338	1.403	1.469	1.539	1.611
6	1.062	1.126	1.194	1.265	1.340	1.419	1.501	1.587	1.677	1.772
7	1.072	1.149	1.230	1.316	1.407	1.504	1.606	1.714	1.828	1.949
8	1.083	1.172	1.267	1.369	1.477	1.594	1.718	1.851	1.993	2.144
9	1.094	1.195	1.305	1.423	1.551	1.689	1.838	1.999	2.172	2.358
10	1.105	1.219	1.344	1.480	1.629	1.791	1.967	2.159	2.367	2.594
11	1.116	1.243	1.384	1.539	1.710	1.898	2.105	2.332	2.580	2.853
12	1.127	1.268	1.426	1.601	1.796	2.012	2.252	2.518	2.813	3.138
13	1.138	1.294	1.469	1.665	1.886	2.133	2.410	2.720	3.066	3.452
14	1.149	1.319	1.513	1.732	1.980	2.261	2.579	2.937	3.342	3.797
15	1.161	1.346	1.558	1.801	2.079	2.397	2.759	3.172	3.642	4.177
16	1.173	1.373	1.605	1.873	2.183	2.540	2.952	3.426	3.970	4.595
17	1.184	1.400	1.653	1.948	2.292	2.693	3.159	3.700	4.328	5.054
18	1.196	1.428	1.702	2.026	2.407	2.854	3.380	3.996	4.717	5.560
19	1.20	1.457	1.754	2.107	2.527	3.026	3.617	4.316	5.142	6.116
20	1.220	1.486	1.806	2.191	2.653	3.207	3.870	4.661	5.604	6.727
21	1.232	1.516	1.860	2.279	2.786	3.400	4.141	5.034	6.109	7.400
22	1.245	1.546	1.916	2.370	2.925	3.604	4.430	5.437	6.659	8.140
23	1.257	1.577	1.974	2.465	3.072	3.820	4.741	5.871	7.258	8.954
24	1.270	1.608	2.033	2.563	3.225	4.049	5.072	6.341	7.911	9.850
25	1.282	1.641	2.094	2.666	3.386	4.292	5.427	6.848	8.623	10.835
26	1.295	1.673	2.157	2.772	3.556	4.549	5.807	7.396	9.399	11.918
27	1.308	1.707	2.221	2.883	3.733	4.822	6.214	7.988	10.245	13.110
28	1.321	1.741	2.288	2.999	3.920	5.112	6.649	8.627	11.167	14.421
29	1.335	1.776	2.357	3.119	4.116	5.418	7.114	9.317	12.172	15.863
30	1.348	1.811	2.427	3.243	4.322	5.743	7.612	10.063	13.268	17.449
35	1.417	2.000	2.814	3.946	5.516	7.686	10.677	14.785	20.414	28.102
40	1.489	2.208	3.262	4.801	7.040	10.286	14.974	21.725	31.409	45.259
45	1.565	2.438	3.782	5.841	8.985	13.765	21.002	31.920	48.327	72.890
50	1.645	2.692	4.384	7.107	11.467	18.420	29.457	46.902	74.358	117.391
55	1.729	2.972	5.082	8.646	14.636	24.650	41.315	68.914	114.408	189.059
60	1.817	3.281	5.892	10.520	18.679	32.988	57.946	101.257	176.031	304.482

FIGURE 83 Here's what $1.00 is worth over time. Just compare the desired interest rate with the desired time period.

often publish growth rates in earnings per share and dividends per share for companies analyzed. Online computer investment database such as www.marketguide.com contain such information.

How Is It Used for Investment Decisions? The growth rate of the company should be compared over the years. In addition, the company's growth rate should be compared to competing companies and industry norms.

The rate of growth in stock price, earnings, and dividends will help determine whether to buy a particular stock. A company with a higher growth rate is theoretically worth more than a company with a low or negative growth rate.

The value of a stock is the present value of all future cash inflows expected to be received by the investor. The cash inflows are dividends and the future sales price.

The growth rate in dividends is a component in Gordon's formula to determine the theoretical value of a stock. If there is a constant growth rate of g every year i.e., $D_1 = D_0 (1 + g)^1$], then the model is

$$P_0 = \frac{D_1}{r - g}$$

where

P_0 equals the current market price of the stock.
D_1 equals the dividends per share in the year I.
r equals the required rate of return.
g equals the constant growth rate in dividends.

Assume that a common stock paid a $3 dividend per share last year and is expected to pay a dividend each year at a growth rate of 10 percent. The investor's required rate of return is 12 percent. The value of the stock would be

$$P_0 = \frac{D_1}{r - g} = \frac{\$3.30}{0.12 - 0.10} = \$165$$

A Word of Caution   When selecting a stock, risk should be considered along with the growth rate.

PRICE-TO-BOOK VALUE RATIO

What Is This Tool?   The ratio compares the market price of a stock to its book value.
How Is It Computed?

$$\frac{\text{Market price per share}}{\text{Book value per share}}$$

The book value per share is equal to the total stockholders' equity divided by total shares outstanding.

Market price per share is based on current prices while book value per share is based on historical cost. Market price per share should typically exceed book value per share because of inflation and good corporate performance over the years.

Assume that a company's market price per share is $30 and its book value per share is $50. The price-to-book value ratio equals

$$\frac{\$30}{\$50} = 0.6$$

This company is not doing well because its market price of stock is 60 percent of its book value. The company's stock has not kept up with increasing prices.

Furthermore, the company's earnings and growth may be deficient. Perhaps, the market is saying that the assets are overvalued.

Where Is It Found? The price-to-book value ratio may be found in financial advisory service publications such as Standard & Poor's. Brokerage research reports sometimes refer to it. If not, the investor can obtain the market price of stock by referring to price quotations in financial newspapers. The book value per share appears in financial advisory publications and, typically, in the company's annual report. Online computer investment databases such as www.marketguide.com contain such information.

How Is It Used and Applied? The analytical implication may be that a company has not been performing well when its market price per share is below its book value per share. The company may be experiencing financial difficulties.

A high price-to-book value ratio is desirable because it shows that the stock market places a higher value on the company. The ratio varies wildly by industry, however.

How Is It Used for Investment Decisions? The investor may have a buying opportunity when book value per share is above market price per share because the stock may be undervalued. On the other hand, if the investor believes the market price per share is too high relative to book value per share, the stock should be sold.

A comparison also should be made to peer groups. (See Figure 84.)

Price-to-Book Value Ratio for Stocks

Company Name	Ticker	Exchange	Price-to-Book 5-Year Avg.
Permian Basin Royalty Trust	PBT	NYSE	20
Sabine Royalty Trust	SBR	NYSE	20
North European Oil Royalty	NET	NYSE	20
Mesabi Trust	MSB	NYSE	20
Avon Products	AVP	NYSE	18.2
General Mills	GIS	NYSE	17.7
Gartner Group	IT	NYSE	17.6
Cambridge Technology Partners	CATP	NASDAQ	17.2
UST, Inc.	UST	NYSE	16.9
Coca-Cola	KO	NYSE	16.6
Presstek	PRST	NASDAQ	16.2
Marine Petroleum Trust	MARPS	NASDAQ	16
LL&E Royalty Trust	LRT	NYSE	14.8
PeopleSoft	PSFT	NASDAQ	14.3
Glaxo Wellcome	GLX	NYSE	13.7
Carlton Communications	CCTVY	NASDAQ	13.5
Coca-Cola Bottling	COKE	NASDAQ	13.4
Total System Services	TSS	NYSE	13.1
Paychex	PAYX	NASDAQ	12.6
Fastenal	FAST	NASDAQ	12.6

FIGURE 84 Here is a look at high price-to-book value ratios (five-year averages) for companies who were profitable each of the three years reported through October 31, 1998. (From Morningstar Inc. With permission.)

A Word of Caution: Other financial ratios have to be considered besides the price-to-book value ratio in selecting a stock. It is just one indicator that should be confirmed by other measures such as the trend in market price per share.

Also See: Share–Price Ratios: Book Value per Share, Share–Price Ratios: Price–Earnings Ratio (Multiple).

PRICE-EARNINGS RATIO (MULTIPLE)

What Is This Tool?   The price–earnings ratio—also known as the *P/E* ratio or the "multiple"—is a company's market price per share divided by its earnings per share. It is an indicator of what traders are willing to pay for a company's profits.

How Is it Computed?

$$\text{Price–earnings ratio} = \frac{\text{Market price per share}}{\text{Earnings per share}}$$

Assume that the market price per share of X Company's stock was $50 and $80, respectively, for 20X1 and 20X2. The earnings per share for those years are $5 and $6, respectively. The price–earnings ratios are computed as

		20X1	20X2
Price–earnings ratio $=$	$\dfrac{\text{Market price per share}}{\text{Earnings per share}}$	$\dfrac{50}{\$5} = 10$	$\dfrac{\$80}{\$6} = 13.3$

The increase in the price–earnings ratio by 33 percent $((13.3 - 10)/10)$ reflects a better perception of the company in the marketplace.

Where Is It Found?   The price–earnings ratios of a company is listed in financial advisory reports (e.g., Standard & Poor's, Moody's, and Value Line) and in the financial pages of daily newspapers (e.g., *Investor's Business Daily, The New York Times,* and *The Wall Street Journal*). Online computer investment databases such as www.marketguide.com contain such information. (See Figure 85.)

How Is It Used and Applied?   An increase in the price–earnings ratio may indicate one or more of the following.

- Investors are more confident in the company. This may be owing to the company's improved financial position (e.g., better cash flow, superior earnings, improved liquidity, and solvency), increase in growth rate, enhanced stability, diversification reducing risk, economic prosperity, new patented products, favorable political environment, and quality management.
- The company's net income may be understated or have a high quality associated with it. This will put a premium on such earnings.

Some companies have high price–earnings multiples that reflect high earnings growth expectations. Young, fast-growing companies often have high price–earnings stocks with multiples over 100.

Price-to-Earnings Ratio for Stocks

Company Name	Ticker	Exchange	P/E Ratio 5-Year Avg.
Presstek	PRST	NASDAQ	130
Energy Research	ERC	AMEX	127
Thermo Cardiosystem	TCA	AMEX	123
Kaiser Ventures	KRSC	NASDAQ	109
Tejon Ranch	TRC	AMEX	105
Airtouch Communications	ATI	NYSE	105
Caere	CAER	NASDAQ	95
Gartner Group	IT	NYSE	95
Nippon Telegraph & Telephone	NTT	NYSE	93
Wind River Systems	WIND	NASDAQ	91
PeopleSoft	PSFT	NASDAQ	91
Parallel Petroleum	PLLL	NASDAQ	91
Noble Affiliates	NBL	NYSE	91
QUALCOMM	QCOM	NASDAQ	89
Alexander's	ALX	NYSE	87
3Com	COMS	NASDAQ	87
Integral Systems	ISYS	NASDAQ	78
Irvine Apartment Communities	IAC	NYSE	78
Cambridge Technology Partners	CATP	NASDAQ	76
Kubota ADR	KUB	NYSE	75

FIGURE 85   Here is a look at high price-to-earnings ratios (five-year averages) for companies who were profitable each of the three years reported through October 31, 1998. (From Morningstar Inc. With permission.)

The price–earnings ratio varies among companies in one industry as well as varying among companies in different industries.

How Is It Used for Investment Decisions?   The price–earnings ratio should be examined by potential investors in deciding whether to invest in the company. A high price–earnings ratio is desirable because it indicates that investors highly value a company's profits. On the other hand, a steady decline in the price–earnings ratio reflects decreasing investor confidence in the business.

Some investors believe that if a company's price–earnings ratio is, relatively, too low, the stock is undervalued and should be bought. On the other hand, some investors believe that if the price–earnings multiple is, relatively, too high, the stock is overvalued and should be sold.

One strategy calls for an investor to buy a company's stock with a low price–earnings ratio and sell the stock when it reaches a high price–earnings ratio. However, this strategy does not always work. A low price–earnings stock may deteriorate further because the business is not doing well.

Other investors watch collective *P/E* ratios of large groups of stocks as an indicator of the strength of the market. So, when in the summer of 1998 the average *P/E* ratio of the stocks in the S&P 500 index was nearing 30, that was seen as trouble. Why? Historically, *P/E* ratios have run around 18 times the S&P 500 earnings. The skeptics were proven correct when in August the market began a horrible tumble that sliced 20 percent off the index in two month's time.

A Word of Caution: The company's price–earnings ratio must be compared over the years and to the price–earnings ratios of competing companies in the industry. Also, many traders now look at the EPS based on Wall Street analysts' consensus estimates for the company's profits in the current or upcoming year. These future or projected *P/E* ratios are extremely important in evaluating companies suffering from losses or at cyclical profit lows rebounding into solid profit years. The price–earnings ratio is only one consideration in making an investment decision. Other factors must be taken into account such as product line, risk, quality of assets, contingencies, and management philosophy.

Also See: Share–Price Ratios: Earnings per Share, Share–Price Ratios: Price–Sales Ratio (PSR).

PRICE-TO-SALES RATIO (PSR)

What Is This Tool?   The price–sales ratio compares the market value of a company's outstanding shares to its sales.

How Is It Computed?

$$PSR = \frac{\text{Market price per share x shares outstanding}}{\text{Sales}}$$

Example: JIL Inc. stock has a market price of $5 per share, outstanding shares are 1.5 million, and sales are $15 million.

$$PSR = \frac{\$5 \times 1.5 \text{ million}}{\$15 \text{ million}} = \frac{\$7.5 \text{ million}}{\$15.0 \text{ million}} = .50$$

Where Is It Found?   PSR information on companies can be found in *Forbes* and *Kiplinger's Personal Finance Magazine*. Online computer investment databases such as www.marketguide.com contain such information. It also can be computed from information contained in the annual report.

How Is It Used for Investment Decisions?   PSR reflects a company's underlying financial strength. A company with a low PSR is more attractive while one with a high PSR is less attractive. As a rule of thumb, investors should avoid stocks with a PSR of 1.5 or more. Further, investors should sell a stock when the PSR is between 3 to 6.

A Word of Caution: The PSR approach to investing is a long-term strategy. Stock should be held for about three to five years because in many cases the investor is dealing with a turnaround situation. (See Figure 86.)

Also See: Share–Price Ratios: Price–Earnings Ratio (Multiple).

Price-to-Sales Ratio for Stocks

Company Name	Ticker	Exchange	Price-to-Sales 5-Year Avg.
Elsevier NV	ENL	NYSE	23.1
Alexander's	ALX	NYSE	22.9
Arizona Land Income	AZL	AMEX	22.4
Presstek	PRST	NASDAQ	20.1
Thermo Cardiosystem	TCA	AMEX	19.7
Reed International	RUK	NYSE	17.2
Shuffle Master	SHFL	NASDAQ	16.4
Theragenics	TGX	NYSE	15.7
Cousins Properties	CUZ	NYSE	14.7
PMC Commercial Trust	PCC	AMEX	14.5
CNS	CNXS	NASDAQ	14.3
Black Hawk Gaming & Development	BHWK	NASDAQ	13.9
Gold Fields of South Africa	GLDFY	NASQ	13.8
Anglo American Gold Investment	AAGIY	NASQ	13.4
RFS Hotel Investors	RFS	NYSE	13.2
Commercial Assets	CAX	AMEX	12.6
San Juan Basin Royalty	SJT	NYSE	12.4
Acorn Holding	AVCC	NASDAQ	12.3
Burlington Resources Coal Seam	BRU	NYSE	12.2

FIGURE 86   Here is a look at high price-to-sales ratios (five-year averages) for companies who were profitable each of the three years reported through October 31, 1998.

TOOL #86

SHARPE'S RISK-ADJUSTED RETURN

What Is This Tool? Some mutual fund analyses use risk-adjusted grades that compare five-year, risk-adjusted returns. This measure is based on a measure developed by Nobel Laureate, William Sharpe. The fund manager is, thus, able to view his excess returns per unit of risk. The index concentrates on total risk as measured by the standard deviation of returns (noted with the Greek letter σ, read as sigma).

How Is It Computed?

$$\text{Sharpe measure} = \frac{\text{Excess returns}}{\text{Fund standard deviation}} = \frac{\text{Total fund return} - \text{Risk-free rate}}{\text{Fund standard deviation}}$$

Example: If a fund has a return of 10 percent, the risk-free rate is 6 percent, and the fund and standard deviation is 18 percent, the Sharpe measure is 0.22.

$$\text{Sharpe measure} = \frac{10\% - 6\%}{18\%} = \frac{4\%}{18\%} = 0.22$$

Where Is It Found? Mutual fund analysis by Morningstar Inc. (www.morningstar.net) and others use it.

How Is It Used for Investment Decisions? In appraising the performance of an investment portfolio, an investor must consider return and risk. The Sharpe index of portfolio performance may be used for this purpose.

An investor should rank the performance of mutual funds based on Sharpe's index of portfolio performance. The funds would be ranked from high to low return. For example, a fund with an index of 0.6 would be far superior to one with an index of 0.3.

Sharpe's index should be compared with other trends as well as with the average market. The larger the index, the better the performance.

A Word of Caution: The index should be used by investors with some mathematical knowledge. Remember, the portfolio with the best risk-adjusted performance will likely not produce the greatest profits. Formulas like Sharpe's take into account the risk undertaken to accomplish profits.

Also See: Performance and Risk: Jensen's Performance Measure (Alpha), Performance and Risk: Treynor's Performance Measure.

TOOL #87

SHORT SELLING

SHORT-INTEREST RATIO (SIR)

What Is This Tool?   Short selling occurs when investors believe that stock prices will drop. Technical analysts look at the number of shares sold short. Short interest measures the number of stocks sold short in the market at any given time which have not yet been repurchased to close out short positions.

How Is It Computed?   The short interest ratio is the latest reported short interest position for the month divided by the daily average trading volume. The SIR is more closely watched than the trading volume of shares sold short.

A high ratio is bullish and a low ratio is bearish. In the past, the ratio for all stocks on the NYSE has hovered between 1.0 and 1.75. A ratio above 1.8 is considered bullish while a ratio below 1.15 is deemed bearish. For example, a ratio of 2 represents 2.0 days of potential buying power. The SIR works best as a bullish indicator after a long-term decline in prices instead of after a long upturn.

Where Is It Found?   The amount of short interest on the NYSE, AMEX, and NASDAQ is published in *Barron's, The New York Times, The Wall Street Journal,* and other financial publications. The exchanges publish short interest figures on about the 20th day of each month.

How Is It Used and Applied?   Looking at short sales is often called a contrary opinion rule. Some believe that an increase in the number of short sellers indicates a bullish market. It is believed that short sellers get emotional and overreact when they are proven wrong and will quickly buy the short-sold stock. Increased short sales and increased market activity will create additional market supply. Then, when the market goes down, the short sellers will buy back their shares, and this will produce increased market demand.

Some believe, however, that increased short selling reflects a downward and technically weak market that results from investors' pessimism. The short seller, in fact, expects a downward market.

Short interest information does have two limitations, however. Some studies have shown that short interest follows the same pattern as market price changes, and data are sometimes not available until two weeks after the short sale occurs.

How Is It Used for Investment Decisions?   By monitoring overall short interest, the investor can foresee future market demand and determine whether the market is optimistic or pessimistic. A very substantial short interest in a single stock should make the investor question the value of the security.

The investor also should examine odd-lot short sales. It is believed that many odd-lotters are uninformed. An odd-lotter short sale ratio of around 0.5 percent indicates optimism; a ratio of 3.0 or more reflects pessimism.

Specialists make markets in securities and are considered "smart money." Investors should watch the ratio of specialists' short sales to the total number of short sales on an exchange. For example, if specialists sell 100,000 shares short in a week and the total number of short sales is 400,000, the specialists' sales constitute 25 percent of all short sales.

Specialists' short sales are a bullish indicator. The specialists keep a book of limit orders on their securities so they are knowledgeable as to market activity at a particular time. However, if most of their short sales are covered, this is a bullish sign. A normal ratio is about 55 percent. A ratio of 65 percent or more is a bearish indicator. A ratio less than 40 percent is bullish.

A Word of Caution: Short-sellers may be right that the market is headed downward.

Also See: Short Selling: Short Sales Position, Short Selling: Specialists/Public Short (*S/P*) Ratio.

SHORT SALES POSITION

What Is This Tool?   Shares are borrowed from a broker in order to sell shares the investor does not own. In short selling, the investor hopes to sell high and to buy back the stock low at a later date. The shares are then returned to the broker. If the stock falls, the investor makes money. If it rises, the investor loses money. It is a speculative form of investment involving considerable risk.

How Is It Computed?

$$\frac{\text{Selling price} - \text{purchase price}}{\text{Gain (or loss)}}$$

where

Selling price equals the shares sold times the selling price per share.
Purchase price equals the shares bought times the cost per share.

Example: Assume that the investor sells short 50 shares of stock with a market price of $25 per share. The broker borrows the shares for the investor and sells them to someone else for $1,250. The brokerage house holds on to the proceeds of the short sale. Later on, the investor buys the stock back at $20 a share, earning a per share profit of $5, or a total of $250.

How Is It Used and Applied?   A short seller must set up a margin account with a stockbroker. If the price suddenly rises, the investor can buy back the stock to minimize his losses. The Federal Reserve requires a short seller to have in a margin account cash or securities worth at least 50 percent of the market value of the stock sold short. Another requirement is that a stock can be sold short only when the stock price has risen. An investor cannot sell short a listed stock that drops steadily from $50 to $30, for example. Short sellers pay brokerage commissions on both the sale and repurchase.

Where Is It Found?   Information on short sales appears in the financial section of newspapers and financial magazines such as *Barron's* and *The Wall Street Journal.*

How Is It Used for Investment Decisions?   The investor may use the following short-selling strategies.

- The investor sells short because he thinks the stock price is going to decline.
- The investor sells short to protect himself if he owns the stock but for some reason cannot sell. If, for example, an investor buys stock through a payroll purchase plan at the end of each quarter but does not get the certificates until several weeks later, it may make sense for him to sell the shares short to lock in the gain.

A Word of Caution: A short seller can incur a significant loss if the stock sold short appreciably increases in market price.

Also See: Short Selling: Short Interest Ratio, Short Selling: Specialists/Public Short (*S/P*) Ratio.

SPECIALISTS/PUBLIC SHORT (SIP) RATIO

What is This Tool?   This ratio is used with the assumption that the speculative public makes an error at market turns and speculators who sell short are among the least astute. What short selling is not done on the exchange by members is by definition done by the public (which may include a few institutions, but they are not really active in short selling). This is one of many so-called smart money rules under which specialists might provide unusual insight into the future.

How Is It Computed?   It is derived by dividing the specialists shorts by the public shorts.

Where Is It Found?   It is found in the Market Laboratory section of *Barron's.*

How Is It Used for Investment Decisions?   It is recommended to smooth the ratio over a four-week average. A ratio of around 3.5 (especially greater than 4.0) is interpreted as a bearish signal because it is a reflection of too much optimism. A *S/P* ratio below about 1.8 (especially near 1.0) indicates considerable pessimism and tends to be bullish.

A Word of Caution: Any indicator can fail at times. Investors should weigh all the evidence.

Tool #**88**

Standard & Poor's 500 Indexes

What Is This Tool? The most widely watched performance benchmark for money managers trying to beat the U.S. stock market.

The best known index for the Standard & Poor's market trackers, it consists of 500 stocks chosen for market size, liquidity, and industry group representation. The S&P 500 is one of the most widely used benchmarks of U.S. equity performance, especially that of large, blue-chip issues. Mutual funds that mimic its performance have become extremely popular investments in the late 1990s. (See Figure 87.)

How Is It Computed? A value-weighted index based on the aggregate market value of the stock (price times number of shares), the indexes are adjusted to include the reinvestment of dividends (cash and stock) and stock splits. A limitation of the index computation is that large capitalized stocks, those having many shares outstanding, heavily influence the index value.

The Standard & Poor's 500 Index

Industry	Number of Companies	Number of Companies as % of 500
Industrials	378	75.6%
Utilities	38	7.6%
Financials	74	14.8%
Transportation	10	2.0%

Exchange Representation

Exchange	Number of Companies	Number of Companies as % of 500
NYSE	457	91.4%
NASDAQ	41	8.2%
AMEX	2	0.4%

Other Statistics

Total Market Value	$8.85 trillion
Mean Market Value	$17.7 billion
Median Market Value	$7.15 billion
Largest Company's Market Value	$284 billion
Smallest Company's Market Value	$451 million

FIGURE 87 Industry Group Representation [as of October 30, 1998]. (Data from Standard & Poor's.)

Where Is It Found?   Almost every daily newspaper, business news TV show, and online financial news service quotes the S&P 500. The *CNBC* cable network runs quotes of trading in S&P 500 index futures in the hours before equity markets open as an indication about how the overall market might go when the opening bell rings on Wall Street. The index also may be accessed through the computerized Standard & Poor's online database at www.advisorinsight.com/pub/indexes/ or on other major Internet financial services such as Bloomberg News at http://www.bloomberg.com/markets/sp500.html, a great tracking of the S&P 500.

How Are They Used for Investment Decisions?   The S&P 500 is widely used to measure institutional performance and gives an indication of total returns on major U.S. equities. The S&P indexes may be used by the investor to gauge market direction and strength. An upward increase in the S&P index is a sign of a bull market while a downward trend is an indication of a bear market. If the overall market is improving, the investor may now view it as a time to buy.

A Word of Caution: While the S&P 500 is seen as the leading benchmark of U.S. stock performance, it is by no means all-inclusive. For example, it is a poor indicator of the performance of small U.S. stocks.

In addition, because so many money managers now attempt to mimic the S&P 500 performance by simply owning those 500 member shares, a stock admittance or expulsion from the index may have a profound impact on trading in that company's shares. That is because money managers must buy or sell that stock to keep pace with the S&P 500.

Also See: Dow Jones Industrial Average, NYSE Indexes, Stock Indexes: Other: Value Line Averages.

Tool #89

Standard & Poor's Other Stock Indexes

What Are These Tools? They are a group of widely watched benchmarks for the U.S. stock market compiled by one of Wall Street's best-known analytical firms.

The S&P Mid-Cap 400-stock index is quickly becoming another popular benchmark. It tracks the performance of 400 medium-sized companies, which some say should, over time, produce superior returns to larger companies because they are perceived to be more managerially nimble. While almost all of the S&P 500 members come from the prestigious NYSE, only 75 percent of the mid-Cap 400 come from the NYSE. While the median market value of an S&P 500 stock is $7 billion, as of October 1998, the median market value of a Mid-Cap 400 stock is $2 billion.

The S&P 100 Index is a condensed version of the S&P 500 and is comprised of 100 highly capitalized stocks. It is largely used as a way to play options listed on the Chicago Board of Trade because, with just 100 members, it is cheaper to trade a basket share representing the index instead of the big S&P 500 roster.

The S&P Small Cap 600-stock index is a barometer for the nation's tiniest stocks. Unlike its larger brethren indexes, the Small Cap 600 is made up of 600 stocks of which roughly one-half come from the NYSE and one-half from the NASDAQ and American markets. The median market value of a Small Cap 600 stock was $22 million as of October 30, 1998.

S&P also has numerous other indexes that track niches of the market. Included in this collection is the S&P Industrials Index, S&P Transportation Index, S&P Financials Index, S&P Utility Index, S&P REIT Index that tracks real estate trusts; and indexes done with Barra Inc. that track growth and value shares by splitting the S&P 500 into two groups based on various financial criteria.

How Are They Computed? The S&P indexes are capitalization (value) weighted to take into account the stock price and the number of outstanding shares. A value-weighted index is based on the aggregate market value of the stock (price times number of shares). The indexes are adjusted to include the reinvestment of dividends (cash and stock) and stock splits. A limitation of the index computation is that large capitalized stocks, those having many shares outstanding, heavily influence the index value.

Where Are They found? The S&P indexes are published by Standard & Poor's. The S&P indexes as well as stock index futures prices can be found in the financial pages of newspapers and business magazines such as *Barron's, The New*

S&P 100	832.18	820.93	827.41	+6.48
S&P 500	1525.30	1502.59	1517.68	+15.09
S&P Midcap	543.97	535.80	542.90	+7.10
S&P Industrials	1854.16	1830.29	1844.92	+14.63
S&P Transportation	603.81	594.70	598.74	+1.41
S&P Utilities	311.49	306.84	310.24	+2.89
S&P Smallcap	223.92	221.37	223.49	+2.12
S&P Financial	158.61	154.31	157.73	+3.42
Major Market Index	1079.93	1069.26	1069.26	-1.92
Value Line Arith	1163.51	1151.75	1160.44	+8.69
Russell 2000	539.14	532.36	537.89	+5.56
Wilshire Smallcap	879.75	862.72	874.69	+11.97
Value Line Geometric	435.39	435.39	435.39	+3.03
Wilshire 5000	14280.04	14280.04	14280.04	+154.46

FIGURE 88   How a newspaper covers the Standard & Poor's stock indexes. (From *Orange County Register* (CA), September 1, 2000, p. B6. With permission.)

York Times, and *The Wall Street Journal.* The indexes also may be accessed through the computerized Standard & Poor's online database at www.advisorinsight.com/pub/indexes/ or on other major electronic financial services such as America Online. (See Figure 88.)

How Are They Used for Investment Decisions?   The performance of respective S&P indexes is an indication of the health of that segment of the stock market. For example, the S&P 400 gives an indication of total returns for mid-sized U.S. stocks.

The investor may buy a stock index futures contract tied into these niche indexes such as the S&P 100 or Mid Cap 400. The latter involves a smaller margin deposit. Stock index futures allow the investor to participate in the general change in the overall stock market. The investor can buy and sell the "market as a whole" rather than specific securities.

A Word of Caution: If the investor anticipates a market move but is unsure which particular stocks will be impacted, he or she could buy a stock index future like those tied to S&P indexes. However, there are high risks involved, so trading in stock index futures should be used only for hedging or speculation.

Also See: Dow Jones Industrial Average, NYSE Indexes, Stock Indexes: Other: Value Line Averages.

Tool #90

Stock Splits

What Are These Tools?   Stock splits occur when a corporation decides to alter its capitalization by either issuing new shares to existing shareholders or, most commonly, will issue new shares at a set rate per share owned. In a reverse stock split, the company reduces the number of shares outstanding. The shares will have the same market value immediately after the reverse split. In other words, the number of shares owned will be less but will be worth more per share.

How Are They Computed?   The numerology of stock splits can be confusing. In a 2-for-1 split, for example, a current shareholder with 100 shares will have 200 shares after the split is completed. As a result, this investor's 100 shares that were priced at $60 each, will now be 200 shares worth $30 a piece. A 3-for-1 split means that 100 shares become 300 (at $20 on a $60 stock) while a 3-for-2 means that 100 shares become 150 (at $40 on that same $60 issue).

On the reverse side, a 1-for-10 split means that a shareholder with 1000 shares will have 100 afterward. The 1000 shares priced at $1 will become 100 shares at $10.

Financial data calculated on a per share basis will need to be recalculated to reflect the changed number of shares and altered share price.

Where Are They Found?   Lists of stock splits are published in many major daily newspapers such as *The Los Angeles Times* and *The Wall Street Journal.*

How Are They Used for Investment Decisions?   Companies say that they make stock splits to keep their share prices in a marketable range—typically between $20 and $50. That keeps a 100-share round lot affordable to many small investors, or so goes the logic.

Reverse splits have become more common as regulators have cracked down on investment schemes involving penny stocks—shares that trade under $5. Such negative publicity has hurt the lure of even the shares of legitimate but very low-priced firms.

A Word of Caution: While many investors like stock splits as a sign that a company believes that its stock price will be rising higher, there is little empirical data to suggest that such splits actually provide any long-term boost to share prices. Nevertheless, they often give a short-term boost to a stock's price and, given the public's approval of the splits, can be viewed as a positive for a company's shares. (See Figure 89.)

Declared stock splits
ADC Telecommunications Inc 2 for 1
Aeroflex Inc 5 for 4
Arca Cp 5 for 4
Certicom Corp g 2 for 1
Diodes Inc 3 for 2
Energis PLC ADS 5 for 1
Patterson Dental 2 for 1
SAP AG 3 for 1
Target Corp 2 for 1
Stock splits last week
Alcoa Inc 2-1
Anaren Microwave 3-2
AT&T Liberty Media A and B 2-1
Chase Manhattan Corp 3-2
Cyber Optics 3-2
Hanover Compressor Co 2-1
Hispanic Broadcasting Corp 2-1
Inter Parfums Inc 3-2
Intl Bancshares Cp 5-4
Investors Financial Svcs 2-1
Juniper Networks 2-1
Metris Cos 3-2
Nexell Therapeutics Inc 1-4 rev
NextLink Communications 2-1
Private Media Group 3-1
Rambus Inc 4-1
Robert Half Intl Inc 2-1
Seacor Smit Inc 3-2
Vishay Intertechnology Inc 3-2

FIGURE 89   How a newspaper covers stock splits. (From *Orange County Register* (CA), June 18, 2000, p. 8. With permission.)

Tool #91

Support and Resistance Levels

What Is This Tool? A support level is the lower end of a trading range; a resistance level is the upper end. At the "support" level, there is support not to have a further decline in price while at the "resistance" level there is resistance to a further price decline. These levels are considered in technical investment analysis.

How Is It Prepared? A chart may be prepared showing the price over time of an individual stock or overall market. The support and resistance prices based on the historical trend are depicted along with the breakout points. Support and resistance lines are actually trend lines drawn through high and low prices.

When prices go to the support level, actual or potential buying in sufficient volume is expected to stop the downward price trend for a sustained period of time. When prices go to the resistance level, actual or potential selling in sufficient volume will halt an upward price trend.

Where Is It Found? A chart for a company or market containing support and resistance levels may appear in brokerage research reports prepared by technical analysts. The investor also can prepare this chart for a company's stock by tracking price over a desired time period.

How Is It Used and Applied? Support may occur when a stock goes to a lower level of trading because new investors may now want to purchase it. If so, new demand will occur in the market. Resistance may take place when a security goes to the high side of the normal trading range. Investors who purchased on an earlier high may view this as a chance to sell the stock at a profit. When market price goes above a resistance point or below a support point (in a "breakout"), investors assume that the stock is trading in a new range and that higher or lower trading volume are imminent.

How Is It Used for Investment Decisions? If the company's stock price is within the support and resistance levels, the stock should be kept. If the stock penetrates the resistance line, there is a buying opportunity. If the support line is broken, there is a selling situation. The investor also should consider volume in addition to price. If a breakthrough in the support or resistance levels is accompanied by heavy volume, this is a stronger indicator. (See Figure 90.)

A Word of Caution: A stock's market price may break through the support or resistance level.

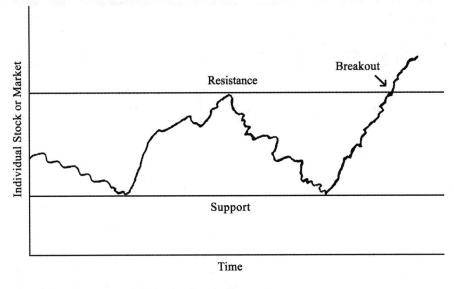

FIGURE 90 Support and Resistance Levels Chart.

Tool #92

Tick and Closing Tick

What Is This Tool?   A tick is a measure of movement in closing stock prices; a positive (+) tick means prices were rising at the end of the day, while a negative (−) tick means prices were falling.

Tick closing prices provide insight into how strong the market was near the close. These tick statistics show the number of stocks whose last price change was an increase, less those whose last move was a down tick.

How Is It Computed?   The closing tick nets all stocks whose last trade was higher than the previous trade (+) on the exchange against all stocks whose last trade was lower (−). It is computed for the NYSE, the AMEX, and for the 30 stocks in the Dow Jones Industrial Average.

Example: A closing tick of +41, for example, means 41 more stocks were rising than falling at their last trade.

Where Is It Found?   It is found in *Barron's*. TV networks such as *CNBC,* The Business Channel, and Nightly Business Report reports closing ticks on a daily basis.

How Is It Used for Investment Decisions?   High positive figures indicate strength while negative ones indicate weakness. In general, the object is to look for positive or negative trends. For example, if the reading went from a negative reading, such as −100, to a positive reading, such as +200, it would be construed a short-term bullish sign. Note, however, extremely high readings supposedly signify a short-term overbought condition and signal a reversal to the downside. By the same token, extremely low tick readings could indicate a short-term oversold condition and are viewed as bullish in the near term.

A Word of Caution: As indicated earlier, TICK readings need to be interpreted with extreme caution because it is not an exact science.

Also See: Arms Index (Tick and Trin).

Tool #93

Total Return

What Is This Tool? Total return is the most complete measure of an investment's profitability. It is based on both price changes of the assets plus dividends or interest payments made by the security's issuer. By including appreciation and income, total return allows for fair evaluations of disparate assets.

How Is It Computed? Total return reflects the price of an investment or portfolio at the start of a period and at the end plus any cash payouts the investment may have generated. Often the calculation assumes that dividends and/or interest payments were immediately reinvested into more shares of that investment.

Examples:

- *For a stock investment*—an investor bought a stock for $5000 plus a brokerage commission of $100. After holding it for one year, during which time he received a dividend of $260, he sold it for $5500 less commissions of $150. The table below shows how to figure the total return.

Net proceeds	($5,500 − $150)	$5,350
Minus net cost	($5,000 + $100)	−$5,100
Plus dividend income		$260
Total return		$510
As a percentage of initial investment	($510/$5,100)	10 percent

- *For a bond investment*—An investor bought a bond for $9900 plus a brokerage commission of $100. After holding it for one year, during which time he received interest at an 8.08 percent rate, he sold it for $9500 less commissions of $100. Here is how to figure the total return.

Net proceeds	($9,500 − $100)	$9,400
Minus net cost	($9,900 + $100)	−$10,000
Plus dividend income		+$800
Total return		$200
As a percentage of initial investment	($200/$10,000)	2 percent

- *For a mutual fund investment*—An investor bought 2000 shares of a no-load fund at $10. After three months, he received $400 in income distributions and $700 in capital gains distributions that were automatically reinvested at

$11, giving him another 100 shares. He or she received the same distribution six month later, only this time he took them in cash. Three months later, he sold the 2100 shares at $12. Here is how to figure the total return.

Net proceeds	(2,100 × $12)	$25,200
Minus net cost	(2,000 × $10)	−$20,000
Plus unreinvested distributions		+$1,100
Total return		$6,300
As a percentage of initial investment	($6,300/$20,000)	31 percent

Where Is It Found?   Total return is discussed in many forms, although most commonly it is used among bond traders and when comparing mutual funds.

In bond markets, interest payments can be wiped out by bad trading, so total return is key. Total return is less frequently talked about when the issue is stocks because dividends are a smaller part of an investor's expected returns.

Services like Lipper Analytical Services of New Jersey and Morningstar of Chicago publish total return figures for funds regularly.

Many stock and bond indexes are calculated on a total return basis. This enables the investor to see the combined effect of appreciation and payouts in such investment benchmarks.

How Is It Used for Investment Decisions?   Total return is a way to produce a level playing field to compare various investments. It is a simple way to check an investment's performance. Over longer periods of time, total return figures also help investors learn the power of compound interest. Simply looking at price appreciation may not adequately reflect an investment's profitability, especially if dividends are routinely reinvested. See Figure 91.

Also See: Lipper Mutual Fund Rankings, Morningstar Mutual Fund Rankings.

Total Return for Large Mutual Funds

Fund Name	5-Year Total Return (Annualized)	10-Year Total Return (Annualized)
Fidelity Magellan	14.7%	17.4%
Vanguard Index 500	19.8%	17.1%
Washington Mutual Investors	18.9%	16.5%
Investment Comp of America	17.0%	15.6%
Fidelity Growth & Income	18.6%	18.3%
Fidelity Contrafund	16.6%	21.5%
Vanguard/Windsor II	18.0%	16.0%
Vanguard/Wellington	15.0%	13.7%
Fidelity Puritan	13.2%	13.6%
American Century (20th) Ultra	14.6%	21.6%
Income Fund of America	12.7%	13.2%
Fidelity Advisors Growth Opportunity	17.7%	18.1%
EuroPacific Growth	9.9%	12.4%
Fidelity Equity-Income	15.8%	14.2%
Vanguard/Windsor	12.4%	12.1%
New Perspective	14.2%	14.1%
Vanguard Institutional Index	19.9%	nh
Putnam Growth & Income A	15.8%	14.9%
Janus Fund	15.7%	17.6%
Pimco Total Return	8.0%	10.4%
Fidelity Equity-Income II	15.6%	nh
MSDW Dividend Growth B	15.7%	15.0%
Putnam Growth & Income B	14.9%	nh
Fidelity Blue Chip Growth	18.1%	20.7%
Franklin California Tax-Free Income	6.3%	7.9%
Growth Fund of America	14.9%	15.3%
Putnam Voyager A	15.5%	17.9%
Templeton Foreign I	6.0%	10.0%
Janus Worldwide	17.8%	nh
Templeton Growth I	10.5%	12.0%
Fundamental Investors	16.7%	15.9%
Fidelity Spartan U.S. Index	19.6%	17.0%
AIM Constellation A	12.0%	18.0%
T. Rowe Price Equity-Income	17.0%	14.2%
Fidelity Asset Manager	10.7%	nh
Vanguard U.S. Growth	21.5%	18.5%
IDS New Dimensions A	16.4%	18.4%
Vanguard/Primecap	19.8%	17.0%
American Mutual	14.6%	14.0%
Vanguard GNMA	7.4%	9.1%
Putnam New Opportunities A	15.7%	nh
Fidelity Growth Company	15.2%	18.7%
Templeton World I	12.4%	11.8%
Bond Fund of America	6.4%	9.1%
T. Rowe Price Int'l Stock	7.7%	9.7%
Janus Twenty	22.3%	23.1%
Franklin U.S. Government Securities I	6.8%	8.5%
Capital Income Builder	13.6%	14.0%
Smallcap World	8.1%	nh
Fidelity Low-Priced Stock	15.1%	nh

FIGURE 91 Here Is a look at total return (price appreciation plus dividends reinvested) for the 25 largest mutual funds based on assets under management. (Funds with "nh" in the 10-year column have no history for that time period). (From Morningstar Inc. With permission.)

Tool #94

Trading Volume

What Is This Tool?   Trading volume is the number of shares traded on a stock exchange or for an individual security for a specified period of time, usually daily.

How Is It Computed?   A tabulation is made of the number of shares transacted for the day.

Where Is It Found?   Trading volume of the overall market and individual stocks can be found in the financial pages of newspapers (e.g., *Barron's, Investor's Business Daily,* and *The Wall Street Journal*) and other financial publications. Program trading activity can be found in *The Wall Street Journal* based on information furnished by the NYSE.

How Is It Used and Applied?   Trading volume trends indicate the health of the market. Price follows volume. For example, increased price can be expected on increased volume.

Market volume of stocks is based on supply–demand relationships. Real and psychological factors influence stock buyers and sellers. A strong market exists when volume increases as prices rise. The market is weak when volume increases as prices decline.

If the supply of new stock offerings exceeds the demand, stock prices will decrease. If the demand exceeds the supply of new stock offerings, stock prices will increase. Supply–demand analysis is concerned more with the short term than with the long term.

Volume is closely related to stock price change. A bullish market exists when there is a new high on heavy trading volume. A new high with light volume, however, is viewed as a temporary situation. A new low with light volume is considered much better than one with high volume because fewer investors are involved. If there is high volume with the new low price, a very bearish situation may exist.

When price goes to a new high on increased volume, a potential reversal may occur where the current volume is less than the prior rally's volume. A rally with declining volume is questionable and may foreshadow a reversal in price. A bullish indicator exists when prices increase after a long decline and then reach a level equal to or greater than the preceding trough. It is a bullish indicator when volume on the secondary trough is less than the first one. When price declines on heavy volume, a bearish indicator exists pointing to a reversal in the trend.

A selling climax takes place when prices decrease for a long period at an increased rate coupled with increased volume. After the selling climax, prices are expected to go up, and the low at the point of climax is not expected to be violated for a long time. A selling climax often occurs at the end of a bear market.

When prices have been rising for several months, a low price increase coupled with high volume is a bearish sign.

An upside–downside index illustrates the difference between stock volume advancing and decreasing and is usually based on a 10-day or 30-day moving average. The index is helpful in predicting market turning points. A bull market continues only where buying pressures remain strong.

An exhaustion move is the last stage of a *major* rise in stock price. It occurs when trading volume and prices drop rapidly. It usually points to a trend reversal.

When net volume increases, accumulation is occurring. When net volume decreases, distribution is taking place. When the net volume line increases or stays constant while the price drops, accumulation under weakness is occurring and a reversal is anticipated. On the other hand, a decrease or constant net volume during a price rise indicates distribution under strength and an impending reversal.

How Is It Used for Investment Decisions? The investor should consider increasing price of a stock on heavy volume to be a much more positive sign than increasing price on light volume. Similarly, decreasing price on heavy volume is a more ominous sign than decreasing price on low volume. Tracking volume indicates the strength or weakness of the security. For example, the investor may have a buying opportunity when a company's stock price begins to rise with increasing volume. This may indicate that the investment community looks favorably upon the stock.

If program trading accounts for a high percentage of share volume, it means that institutional investors (e.g., brokerage firms) are very active. This may have a pronounced effect on the price of stock in either direction. The investor must realize that when a stock is widely held by institutions, price may vary greatly as the institutions buy or sell. The investor should try to determine why the company's stock is being bought or sold significantly by institutions. What do the institutions know?

The investor also should consider secondary distributions, which are the number of new stock issues by already public companies. A high number indicates an overheated market.

A Word of Caution: Trading volume does not always correlate to the magnitude of stock price change. For example, trading volume may be high but the Standard & Poor's 500 index may change only modestly by one or two points.

Also See: Breadth (Advance–Decline) Index, Trading Volume Gauges: Speculation Index.

Tool #95

Trading Volume Gauges

LOW-PRICE ACTIVITY RATIO

What Is This Tool? The ratio compares high risk (speculative) stocks to blue chip (low risk) stocks.

How Is It Computed?

$$\frac{\text{Volume of low-priced speculative securities}}{\text{Volume of high-quality securities}}$$

Example: The volume of low-priced speculative stocks and high-quality stocks are 15 million shares and 140 million shares, respectively. The ratio is 10.7 percent.

Where Is It Found? This measure is published in *Barron's*.

How Is It Used for Investment Decisions? The market top may be indicated when the volume of low-priced securities increases compared to the volume of blue chip stocks. A ratio of less than about 3 percent indicates a market bottom, which is the time to buy because increasing stock prices are expected. However, a ratio about 7.6 percent indicates a market peak, which is the time to sell securities.

A Word of Caution: The investor's definition of a blue chip or speculative stock is subjective. What is speculative to one person may not be for another person. (See Figure 92.)

NET MEMBER BUY–SELL RATIO

What Is This Tool? The ratio is the volume of shares purchased compared to the volume of shares sold by members of the stock exchange. These members include specialists in securities and floor traders.

How Is It Computed?

$$\frac{\text{Volume of securities bought by members}}{\text{Volume of securities sold by members}}$$

Example: If shares bought and sold by members of the stock exchange are 60 million and 45 million, respectively, the ratio is 1.33.

Where Is It Found? The net member buy–sell statistic may be found in *Barron's*.

NEW YORK STOCK EXCHANGE DATA BANK

	Latest Month	Preceding Month	Year-Ago Month
Stock Volume (Thous. shs) July	19,076,891	21,703,321	15,359,961
Bond Volume (Thous. $) July	160,465	194,998	230,172
Warrant Volume (Thous. shs) July	3,220	2,690	3,793

Security Sales

	Latest Month	Preceding Month	Year-Ago Month
All registered exchanges (Mil $) Nov.	869,433.6	973,300.8	693,629.9
Value listed stocks (Mil $) July	12,220,500	11,871,200	11,980,300
Value shares transacted (Mil $) July	799,090	918,689	687,126
Customers' margin debt (Mil $) July	244,970	247,200	178,360
Customers' credit balances (Mil $) July	71,730	64,970	44,330
Free credit bal. cash accounts (Mil $) July	74,970	74,140	60,000
Short interest (Thous. shs) Aug.	4,182,377	4,231,986	3,764,783
Short interest ratio (d) Aug.	4.40	4.30	4.00

NYSE Seat Sales

	Latest Month	Preceding Month	Year-Ago Month
June 23(Thous.$)	1,600	1,700	2,500

(d) For period June 13-to-July 14. p-Preliminary. r-Revised.

FIGURE 92   How a newspaper covers trading activity. (From *Barron's Market Week,* August 28, 2000, p. 78. With permission.)

How Is It Used for Investment Decisions?   Members of the stock exchange are considered "smart money" who have significant expertise in securities. If they are net buyers (buyers minus sellers), this is a bullish sign. If they are net sellers, this has a bearish connotation.

The trend over time in member activity is a reflection of the direction of market confidence.

A Word of Caution: The specialists and floor traders in a stock may be wrong in their investment decisions. They are not infallible. For example, they did not expect the stock market crash in October 1987.

ODD-LOT THEORY

What Is This Tool?   An odd lot is a transaction involving fewer than 100 shares of a security. It is usually done by small investors. Odd-lot trading reflects popular opinion. The odd-lot theory rests on the rule of contrary opinion. In other words, an investor determines what others are doing and then does the opposite.

How Is It Computed?   An odd-lot index is a ratio of odd-lot purchases to odd-lot sales. This ratio usually is between 0.40 and 1.60. The investor also may look at the ratio of odd-lot short sales to total odd-lot sales, and the ratio of total odd-lot volume (buys and sells) to round-lot volume (units of 100 shares each) on the NYSE. These figures serve to substantiate the conclusions reached by the investor in analyzing the ratio of odd-lot selling volume to odd-lot buying volume. Figure 93 shows a chart of the ratio of odd-lot purchases to sales (plotted inversely).

$$\text{Odd-lot short ratio} = \frac{\text{Odd-lot short sales}}{\text{Average of odd-lot purchases and sales}}$$

Example: In one period, odd-lot purchases were 1,500,000 shares while odd-lot sales were 3,000,000 shares. The odd-lot index is therefore 0.50. The index last period was 1.2. The investor should now buy securities because odd-lot traders, who reflect popular opinion, are selling.

Where Is It Found?   Odd-lot trading data are published in *Barron's, Investor's Business Daily, The New York Times,* and *The Wall Street Journal.* Volume is typically expressed in number of shares instead of dollars. The *Securities and Exchange Commission Statistical Bulletin,* another source of data, refers to volume in dollars. Figure 93 shows odd-lot trading as published in *The Wall Street Journal.*

How Is It Used and Applied?   According to the odd-lot theory, the small trader is right most of the time but misses key market turns. For example, odd-lot traders correctly start selling off part of their portfolios in an up market trend but, as the market continues to rise, the small traders try to make a killing by becoming significant net buyers. This precedes a market fall. Similarly, it is assumed that odd-lotters will start selling off strong prior to a bottoming of a bear market. When odd-lot volume rises in an increasing stock market, the market is about to turn around.

FIGURE 93   How a newspaper covers odd-lot activity. (From *The Wall Street Journal,* August 25, 2000, p. C6. With permission.)

How Is It Used for Investment Decisions?   The investor should buy when small traders are selling. Similarly, the investor should sell when small traders are buying.

A Word of Caution: Stock market research does not fully support the odd-lot theory.

UP-TO-DOWN VOLUME RATIO

What Is This Tool?   The up-to-down ratio looks at the number of advancing issues relative to declining ones. It is used in technical investment analysis.

How Is It Computed?   The upside–downside index is the ratio of stock volume advancing to stock volume decreasing. It is usually based on a 10-day or 30-day moving average.

Where Is It Found?   The up-to-down volume ratio may be found in *Barron's* and *The Wall Street Journal.*

How Is It Used for Investment Decisions?   The ratio aids in determining whether accumulation or distribution is occurring. It is helpful in predicting market turning points. For example, a bull market continues only where buying pressures remain strong.

A Word of Caution: Although the up-to-down ratio is increasing, the market still may be headed for future falloff in stock prices owing to an overvalued situation.

Also See: Breadth (Advance–Decline) Index.

Tool #96

Treynor's Performance Measure

What Is This Tool? The index can be used to measure portfolio performance. It is concerned with systematic (beta) risk.

How Is It Computed?

$$T_p = \frac{\text{Risk premium}}{\text{Portfolio's beta coefficient}}$$

Example: An investor wants to rank the two stock mutual funds he or she owns. The risk-free interest rate is 6 percent. Information for each fund follows.

Growth Fund	Return	Fund's Beta
A	14 percent	1.10
B	12 percent	1.30

$$T_A = \frac{14\% - 6\%}{1.10} = 7.27 \text{ (first)}$$

$$T_B = \frac{12\% - 6\%}{1.30} = 4.62 \text{ (second)}$$

Fund A is ranked first because it has a higher return relative to Fund B.

Where Is It Found? The index can be computed based on information obtained from financial newspapers such as *Barron's* and *The Wall Street Journal.*

How Is It Used for Investment Decisions? The index can be used to rank mutual fund investments in terms of return performance considering risk.

Also See: Sharpe's Risk-Adjusted Return, Performance and Risk: Jensen's Performance Measure (Alpha).

Tool #97

Value Averaging

What Is This Tool? Value averaging is an investment strategy developed by Harvard professor Michael E. Edleson. The premise of value averaging is to make the value of an investor's stock holdings increase by $1000 (or some other preset amount) each investment period instead of investing a fixed dollar amount as is done with dollar-cost averaging. This way, investors will realize higher returns at lower per share prices. Edleson claims that this strategy beats dollar-cost averaging about 90 percent of the time.

How Is It Computed? With dollar-cost averaging, an investor automatically buys more shares when prices are low and fewer shares when prices are high. Buying is all the investor does, however. With value averaging, an investor would sell part of his investment if the value goes up too high.

Example: If, after four months, our investor's investment was worth $5000, he would sell $1000 worth of shares the fifth month to bring the account value down to $4000. In other words, value averaging forces him to sell high.

How Is It Used for Investment Decisions? There is no decision involved. Value averaging is an investment strategy that takes emotions out of the investment process. As with dollar-cost averaging, an investor can invest without worrying whether this is the right time to buy or sell.

A Word of Caution: Value averaging works well for a fund that stays fully invested in stocks, such as one that tracks the S&P's 500 index. In fact, it works best with no-load index funds that charge no commission to buy or sell. Value averaging requires more work and will probably result in more taxable transactions.

Also See: Dollar-Cost Averaging.

TOOL #98

VALUE LINE AVERAGES

What Are These Tools?   The Value Line Composite Index is a broad-based measure of prices of NYSE, AMEX, and NASDAQ blue chip and second-tier stocks; it represents approximately 95 percent of the market value of all U.S. securities. It was developed and has been maintained, and published since 1961 by Arnold Bernhard and Co., which was renamed Value Line, Inc., upon its reorganization in October 1982.

To be included, a stock must

1. Have a reasonable market value, or capitalization.
2. Have a strong trading volume, which is a measure of investor interest.
3. Have a high degree of investor interest, as represented by the number of requests for information on a specific stock by the subscribers of the Value Line Investment Survey.

Component stocks are rarely dropped from the index; if one is, it is because the company has

1. Gone bankrupt with little hope of revitalization and continued investor interest.
2. Merged with another company.
3. Gone private.

How Are They Computed?   From 1961 until 1988, the "Value Line Index," as it is commonly known, was computed only as a price-weighted geometric average of the approximately 1700 prices, with a base level of 100 as of June 30, 1961. Each stock price, therefore, has the same weight. Many analysts feel that the use of a geometric average imparts a downward bias to the index, so that it climbs slower and declines faster than a measure based on an arithmetic average. Because of the inclusion of many second-tier issues, the index is also much more volatile than most broad-based indexes.

In March 1988, Value Line began publishing a second price index called the Value Line Arithmetic Index. The only difference between the Value Line Composite and Arithmetic indexes is the method used in calculating them.

Besides the geometric and arithmetic averages, Value Line maintains three lesser-known price averages presenting industrial, railroad, and utility issues.

The Value Line Composite Index was the basis for the first cash-settled tradable stock index contract, initially traded on February 24, 1982, on the Kansas City Board of Trade (KCBT). Here, "maxi" and "mini" index futures contracts are still traded on the KCBT under the ticker symbols KV and MV, respectively. In addition, similar stock index option contracts began trading January 11, 1985, on the Philadelphia Exchange under the ticker symbol XVL.

Where Are They Found?   The indexes are published by Value Line. The Geometric Index and the Arithmetic Index appear in *Barron's*. Information also may be obtained from online services.

How Are They Used for Investment Decisions?   The Value Line Averages are a basis to evaluate the performance of the stock market and certain segments within. Some investors use these indexes because they more closely correspond to the variety of stocks that small investors may have in their portfolios. A bullish market may represent a time to buy if the peak has not been reached. On the other hand, a bear market may be a time to unload securities if the bottom has a way to go.

VALUE LINE CONVERTIBLE INDEXES

What Are These Tools?   The indexes measure the performance of convertible bonds and preferred stocks. Convertible securities can be converted (exchanged) into common stock at a later date.

How Are They Computed?   The index is equally weighted and measures the price performance of about 575 convertible securities including preferred, bonds, and Euro-convertibles. The Value Line Total Returns Index includes the income on the issues. The Value Line Warrant Index is equally weighted and measures price performance of 85 warrants. The base, as of March 1, 1982, is 100. Both are computed by Value Line.

Where Are They Found?   The indexes are found in financial newspapers such as *Barron's.*

How Are They Used for Investment Decisions?   The investor examines tile indexes to see how well convertibles are doing in price and yield. If the investor believes that convertible security prices are undervalued, a buying opportunity exists.

Also See: Lipper Mutual Fund Indexes.

Tool #99

Wilshire 5000 Equity Index

What Is This Tool?   The Wilshire 5000 is the broadest weighted index of all common stock issues on the NYSE, AMEX, and the most active issues on the over-the-counter market. Approximately 85 percent of the securities are traded on the NYSE. The index's value is in billions of dollars. It includes about 6000 stocks (not 5000 as its name would suggest) so it is representative of the overall market.

How Is It Computed?   The stocks included in the index are weighted by market value. It covers total prices of all stocks with daily quotations. The base period is December 31, 1980, at which time the total market value of all stocks in the index was $1404.9 billion, or $1.4 trillion. By the middle of 1998, it had risen to $11,106 billion, or $11 trillion.

Where Is It Found?   The Wilshire index is published by Wilshire Associates (Santa Monica, CA). The index is reported weekly in *Barron's* and in each issue of *Forbes.* It is reported daily in many newspapers such as *Investor's Business Daily, The New York Times,* and *The Wall Street Journal.*

How Is It Used for Investment Decisions?   The investor should use the Wilshire 5000 index as a barometer of the overall stock market condition. In a bull market the index will be increasing, but in a bear market the index will be decreasing. A stock may be bought if the index is on an upward move and the investor feels that prices will continue moving up. However, if stock conditions are deteriorating as evidenced by a declining index, the investor should sell the stock if he believes that conditions will worsen. (See Figure 94.)

A Word of Caution: The equity index may have increased in a day but a component element of the index (e.g., AMEX issues or NASDAQ issues) may have decreased.

YAHOO!FINANCE

| Dow 10824.80 +124.67 (+1.17%) | Nasdaq 3592.92 +24.02 (+0.67%) | S&P 500 1451.41 +15.18 (+1.06%) |
| NYSE Volume 785,000,000 | Russell 2000 512.86 +1.19 (+0.23%) | 30-Yr Bond 5.936% +0.011 |

[   ] Get Quotes [Detailed   ▼] symbol lookup

Welcome, jonlansner   My Yahoo! View - Customize - Sign Out

Accounts [add: bank, credit cards]

Portfolios [manage - create - add: brokerage, 401(k), mutual fund]
oink - OC guys - OC BBs - unOC - dj - foley - grillings - OC junk | Java Mgr

Quotes   DATEK ONLINE ▷   ndb.com   TD WATERHOUSE   E★TRADE

Need a mortgage? Visit the Loan Center for quotes, rates, recommendations and more.   Click to trade or open an account. - Important Disclaimer

Views: Quik - DayWatch - Performance - Snapper - short - name only - mkt cap - Real-time ECN NEW! - Detailed -
[Create New View]

Tue, October 3 2000 2:34pm ET - U.S. Markets close in 1 hour and 26 minutes.

Wilshire 5000 (:^WIL5) - More Info: N/A					
Last Trade Oct 2 · **13522.25**	Change -91.09 (-0.67%)	Prev Cls 13613.34	Volume N/A	Div Date N/A	
Day's Range 13522.25 - 13522.25	Bid N/A	Ask N/A	Open 13522.25	Avg Vol N/A	Ex-Div N/A
52-week Range 12474.650 - 14751.640	Earn/Shr N/A	P/E N/A	Mkt Cap N/A	Div/Shr N/A	Yield N/A

WIL5 @ 2:42pm ©Yahoo!
13800
13750   X
13700
13650 - - - - - - - - - -
13600
13550
3-Jul 10am 12pm 2pm 4pm 6pm
Small: [1d | 5d | none]
Big: [1d | 5d]

Add to My Portfolio - Set Alert   Non-Tables Version - Download Spreadsheet

FIGURE 94   How a web site covers the Wilshire 5000 index.(Reproduced with permission of Yahoo! Inc.© 2000 by Yahoo! Inc. YAHOO! and the YAHOO! Logo are trademarks of Yahoo! Inc.)

TOOL #**100**

YIELD CURVE

What Is This Tool?   While basically a graphic representation of bond yields vs. bond maturities, the yield curve's shape is often quoted by analysts as a description of the bond market's condition. The curve breaks bonds into three categories by maturity—short (less than 5 years in length), intermediate (5 to 10 years), and long (10-plus years).

How Is It Computed?   The curve is drawn by graphing the various maturities of one type of bond, from shortest to longest (typically from overnight to 30 years), against the yields those bond maturities are currently producing. The curve is then sometimes compared to yield curves of other securities or of the same securities at earlier times.

Where Is It Found?   On a daily basis, readers can find the yield curve in *Investor's Daily* and *The Wall Street Journal*. Many other publications and newsletters, such as the *Bond Fund Report* or *Grant's Interest Rate Observer,* as well as reports from investment houses, typically discuss and chart the yield curve. Further, the Bloomberg Web site (www.bloomberg.com/markets/C13.html), for example, presents the Treasury yield curve, as shown in Figure 95.

How Is It Used for Investment Decisions?   The slope of the yield curve and relative changes in the shape can help investors understand bond market conditions.

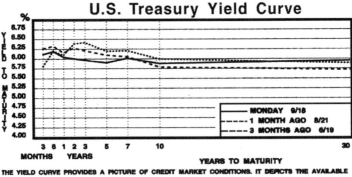

FIGURE 95   Here's how a newspaper would graph the yield curve. (From *Investor's Business Daily,* September 19, 2000, p. B22. With permission.)

When the yield curve is steep—short-term rates low, long-term rates high, which is considered the normal shape—rate-seeking investors must take on the added risk of owning long-term bonds to boost their returns. Long-term bond prices are most volatile because their fixed payouts can be dramatically hurt if, for instance, inflation were to rise and erode that income stream's buying power.

An inverted yield curve creates a difficult choice for investors. Should lower long-term rates be locked in? Or can investors take the chance that the currently more attractive short-term rates will stay? This often occurs in times of economic difficulty.

Then, there is a flat yield curve when passbook rates might be the same as mortgage rates. This is often seen as a signal of an upcoming dramatic change in the interest rate environment.

A Word of Caution: Yield is not the only consideration when it comes to buying **a** bond. The price risk of holding long-term bonds should be studied. Credit quality is another important factor—a high current yield is not a valuable holding after an issuer defaults.

Also See: Bond Ratings, Duration, Economic Indicators and Bond Yields, Interest Rates: 30-Year Treasury Bonds, Interest Rates: Three-Month Treasury Bills.

Tool #101

Yield on an Investment

ANNUAL PERCENTAGE RATE (EFFECTIVE ANNUAL YIELD)

What Is This Tool?   The annual percentage rate (APR), also known as the effective annual yield, is the true (effective) rate of interest earned on savings that reflects the frequency of compounding.

How Is It Computed?   APR is computed as follows:

$$APR = (1 + r/m)^m - 1.0$$

where

> r equals the stated, nominal, or quoted rate.
> m equals the number of compounding periods per year.

Example: If a bank offers a 6 percent interest, compounded quarterly, the APR is

$$
\begin{aligned}
APR &= (1 + .06/4)^4 - 1.0 \\
&= (1.015)^4 - 1.0 \\
&= 1.0614 - 1.0 \\
&= .0614 \\
&= 6.14 \text{ percent}
\end{aligned}
$$

This means that one bank offering 6 percent with quarterly compounding and another bank offering 6.14 percent with annual compounding would both be paying the same effective rate of interest. See Figure 96.

Where Is It Found?   Banks competing for deposits, such as certificates of deposit (CDs), usually spell out their true (effective) annual yields, along with their nominal interest rates in their advertising, inside bank offices, or disclose them over the phone.

How Is It Used for Investment Decisions?   Different types of investments use different compounding periods. For example, most bonds pay interest semiannually. Some banks pay interest quarterly. If an investor wishes to compare investments with

Effective Interest Rates with Compounding

Nominal Rate	Annual Compounding	Semiannual Compounding	Quarterly Compounding	Monthly Compounding	Daily Compounding
6.00%	6.00%	6.09%	6.14%	6.17%	6.18%
7.00%	7.00%	7.12%	7.19%	7.23%	7.25%
8.00%	8.00%	8.16%	8.24%	8.30%	8.33%
9.00%	9.00%	9.20%	9.31%	9.38%	9.42%
10.00%	10.00%	10.25%	10.38%	10.47%	10.52%
11.00%	11.00%	11.30%	11.46%	11.57%	11.62%
12.00%	12.00%	12.36%	12.55%	12.68%	12.74%

FIGURE 96 Here is a look at how much difference varied frequency of compounding can make on the rate of interest paid at annualized rates.

different compounding periods, he needs to put them on a common basis. APR is used for this purpose.

A Word of Caution: Yield is one thing; risk is another. Furthermore, in the case of CDs, investors should ask about other provisions such as the penalty for early withdrawal. For example, the so-called brokered CDs (high-yield CDs purchased through brokerage firms) allow investors to cash out of CDs without paying an interest penalty.

CURRENT YIELD ON A BOND

What Is This Tool? It is the measure of what the coupon payments on a bond are worth when the market value or last-trade price of the bond, rather than the par value or principal value of the bond, is considered. If a bond is selling at par, its coupon rate and current yield will be the same.

How Is It Computed? A simple formula provides this information.

$$\text{Current yield} = \frac{\text{Coupon payments per year}}{\text{Market price}}$$

Example: A $1000 par value bond pays a 9 percent coupon rate. It is currently selling for $1075.

$$\text{Current yield} = \$90/\$1{,}075 = 0.0837 = 8.37 \text{ percent}$$

However, this simple current yield formula does not take into account the maturity of the bond. To do that, an investor must calculate the yield to maturity, or at least a close approximation, by using the following formula.

$$\text{Yield to maturity} = \frac{\text{Annual coupon payment} + \dfrac{(\text{Face value} - \text{market price})}{\text{Years to maturity}}}{(\text{Face value} + \text{market value})/2}$$

Example: Using the same bond, assume a 10-year maturity.

$$\text{Yield to maturity} = \frac{\$90 + (\$1,000 - \$1,075)/10}{(1,000 + \$1,075)/2}$$

$$= \$82.50/\$1,037.50 = 0.0795 = 7.95 \text{ percent}$$

Where Is It Found?   The typical bond listing in newspapers such as *Barron's, The New York Times,* and *The Wall Street Journal* contains yield to maturity information along with coupon rate and last-trade prices. Computerized securities industry databases, available at most brokerage houses, also contain such data.

How Is It Used for Investment Decisions?   By using the current yield, an investor can more accurately see the value of the income stream being generated by a specific bond. The investor can then properly compare it to other income-producing investments. However, going the extra step and calculating the yield to maturity, the investor can get an even better picture of the value of an income stream. This is particularly important when investors are buying bonds at a premium price above the par value.

A Word of Caution: Yield is not the only decision an investor must consider when buying a bond. Credit quality, the callability (or early payoff) and a bond's sensitivity to interest rate swings are also important factors.

Also See: Bond Ratings, Duration.

CURRENT YIELD ON A STOCK

What Is This Tool?   Current yield on a stock is a way to evaluate the income stream created by a dividend on a common or preferred stock as it relates to the market price of those shares. Dividends are paid to shareholders out of the profits a company makes.

How Is It Computed?   Using a simple formula, investors combine dividend information and share price.

Dividend yield = Dividend paid last 4 quarters/Latest share price

Example: Flood City & Co. pays a quarterly dividend of 25 cents. Its shares last traded for $24.

Dividend yield = ($0.25 + $0.25 + $0.25 + $0.25)/$24 = .0417 = 4.17 percent

Some investors, however, like to calculate the yield of a stock based on their original cost, not current selling prices. To do that, they just substitute the purchase price for the last trade. In the previous example, the investor paid $20 for his shares.

Dividend yield = ($0.25 + $0.25 + $0.25 + $0.25)/$20 = .050 = 5.0 percent

Where Is It Found?   The stock tables of publications, such as *Barron's* and *The Wall Street Journal* contain current yield information on common and preferred stock. Computerized securities industry databases such as Reuters and Bloomberg, available at most brokerage houses, also contain such data.

How Is It Used for Investment: Decisions? The current dividend yield gives an investor the chance to compare the cash-generating ability of income-producing stocks to other investments.

The yield also can be used as a measure of the market's outlook for a particular issue. When a stock yield is near an all-time high, analysts might say either that the stock is underpriced or that the dividend is in danger of being cut. When a stock yield is near a low, the stock may be overvalued or the chances of a dividend increase are remote.

A Word of Caution: Many stocks, particularly so-called growth issues, of younger companies that reinvest their profits back into their business do not pay dividends or pay very small ones. Dividend yield is only one part of the equation when choosing a stock to buy. (See Figure 97.)

Dividend Yield for Stocks

Company Name	Ticker	Exchange	Dividend Yield Current	Past Year's Dividend	Stock Price (10/31/98)
Chrysler	C	NYSE	3.3%	$1.60	$48.13
Ford Motor	F	NYSE	3.2%	$1.69	$54.25
Philip Morris Companies	MO	NYSE	3.2%	$1.64	$51.13
General Motors	GM	NYSE	3.2%	$2.25	$63.19
Bank One	ONE	NYSE	3.0%	$1.48	$48.75
Bell Atlantic	BEL	NYSE	2.9%	$1.57	$53.19
BankAmerica	BAC	NYSE	2.6%	$1.52	$57.50
First Union	FTU	NYSE	2.6%	$1.48	$58.00
Amoco	AN	NYSE	2.6%	$1.48	$56.25
Chase Manhattan	CMB	NYSE	2.4%	$1.34	$56.81
Exxon	XON	NYSE	2.3%	$1.64	$71.63
DuPont	DD	NYSE	2.3%	$1.33	$57.75
Washington Mutual	WAMU	NASDAQ	2.2%	$0.92	$37.44
AT&T	T	NYSE	2.1%	$1.32	$62.50
SBC Communications	SBC	NYSE	2.0%	$0.92	$46.31
American Home Products	AHP	NYSE	1.8%	$0.86	$48.94
Baker Hughes	BHI	NYSE	1.6%	$0.46	$22.06
PepsiCo	PEP	NYSE	1.5%	$0.51	$33.75
Merrill Lynch & Company	MER	NYSE	1.5%	$0.88	$59.00
Boeing	BA	NYSE	1.5%	$0.56	$37.56

FIGURE 97   Here is a look at high dividend yields for actively traded stocks as of October 31, 1998. Dividend yield is calculated by dividing the past year's dividends by the current stock price. (From Morningstar Inc. With permission.)

DIVIDEND PAYOUT RATIO

What Is This Tool?   The dividend payout ratio measures the percentage of net income paid out in dividends.

How Is It Computed?

$$\text{Dividend payout} = \frac{\text{Dividends per share}}{\text{Earnings per hare}}$$

Example   Assume the following.

	20X1	20X2
Cash dividends	$ 200,000	$ 500,000
Net income	1,000,000	1,100,000

The dividend payout ratios are 20 percent in 20X1 and 45.5 percent in 20X2, respectively. The investor would look upon the increase in dividend payout favorably because a higher dividend distribution is typically associated with the company performing better. The investor usually likes to receive more dividends because it is available cash and is associated with less uncertainty about the business.

Where Is It Found?   The dividend payout ratio often appears in financial advisory service publications (e.g., the *Value Line Investment Survey*), brokerage research reports, and *Business Week's* Corporate Scorecard. The investor also can determine it from readily available data in a company's annual report because dividends per share and earnings per share are always presented. Dividends per share of stocks in the Dow Jones Averages is published in *Barron's*. The week's dividend payments also appear in *Barron's*. *Forbes* lists stocks with strong or laggard dividends.

How Is It Used for Investment Decisions?   A decline in dividend payout will cause concern to the investor because fewer earnings are being distributed in the form of dividends. Perhaps the company is running into financial difficulties forcing it to cut back on its dividends. A "red flag" is raised when a company pays out in dividends more than its earnings. See Figure 98.

Some industries, such as utilities, are known for their stable dividend records. This is attractive for an individual such as an elderly investor who relies on fixed income.

A Word of Caution: A company should retain earnings rather than distribute them when the corporate return exceeds the return that investors could obtain on their money elsewhere. Further, if the company obtains **a** return on its profits that exceed the cost of capital, the market price of its stock will be maximized.

Also See: Share–Price Ratios: Book Value per Share, Current Yield on a Stock, Share–Price Ratios: Earnings per Share, Share–Price Ratios: Price–Earnings Ratio (Multiple).

Dividend Payout Ratio

Company Name	Ticker	Exchange	Dividend Payout Ratio	EPS (Trailing 12 Months)	Dividend (Trailing 12 Months)
Boeing	BA	NYSE	311%	$0.18	$0.56
Du Pont	DD	NYSE	102%	$1.30	$1.33
Bell Atlantic	BEL	NYSE	87%	$1.80	$1.57
Amoco	AN	NYSE	81%	$1.83	$1.48
Philip Morris Companies	MO	NYSE	63%	$2.60	$1.64
General Motors	GM	NYSE	60%	$3.78	$2.25
First Union	FTU	NYSE	56%	$2.64	$1.48
Exxon	XON	NYSE	55%	$2.96	$1.64
Baker Hughes	BHI	NYSE	55%	$0.83	$0.46
Bank One	ONE	NYSE	52%	$2.87	$1.48
Gillette	G	NYSE	51%	$0.96	$0.49
BankAmerica	BAC	NYSE	47%	$3.23	$1.52
SBC Communications	SBC	NYSE	46%	$2.01	$0.92
Bristol-Myers Squibb	BMY	NYSE	46%	$3.41	$1.55
Merck	MRK	NYSE	45%	$4.16	$1.89
Warner-Lambert	WLA	NYSE	45%	$1.34	$0.61
Eli Lilly & Company	LLY	NYSE	45%	$1.78	$0.80
General Electric	GE	NYSE	44%	$2.71	$1.20
American Home Products	AHP	NYSE	43%	$2.01	$0.86
PepsiCo	PEP	NYSE	41%	$1.24	$0.51

FIGURE 98 Here is a look at high dividend payout ratios for actively traded stocks as of October 31, 1998. Dividend payout ratio is calculated by dividing the past year's dividends by the past year's earnings per share. A ratio above 100 percent means that a company is paying out more than all of its profits to shareholders in the form of dividends. That is a condition that cannot last too long. (From Morningstar Inc. With permission.)

TAX-EQUIVALENT YIELD

What Is This Tool? The tax-equivalent yield shows what a saver would have to earn on a taxable income-producing investment before taxes to equal tax-free bond payouts.

How Is It Computed? First, an investor must figure his marginal tax rate, that is, what he pays on each extra dollar of income. That can be difficult when finding the marginal rate for double tax-free investments, those free of both state and federal taxes. The formula is

$$\text{Tax-equivalent yield} = \frac{\text{Tax-free return}}{1 - \text{Marginal tax rate}}$$

Example: A municipal bond pays an interest rate of 6 percent. The investor's tax rate is 34 percent. The equivalent rate on a taxable instrument is

$$\frac{0.06}{1 - 0.34} = \frac{0.06}{0.66} = 9.1 \text{ percent}$$

Where Is It Found? Tax-equivalent yield is widely used in marketing and promotional materials for tax-free investments, often in large, dramatic type. Legally, the assumptions used to calculate the tax-equivalent yield must accompany it. They are often found in very small type at the bottom of such sales devices. (See Figure 99.)

How Is It Used for Investment Decisions? It is a way to compare tax-free bond yields to taxable returns. It must be used carefully, however. Experts remind investors that the tax status of an investment should not be the only reason to buy it.

Tax-free investments come in various credit qualities. The investor must use a taxable investment of equal safety—both from a credit quality and maturity perspective—when making such comparisons.

In addition, a saver's tax rate is important. According to Internal Revenue Service (IRS) statistics, 1 in 6 investors with tax-free income does not earn enough to have a tax bill sufficiently large to make tax-free investments outperform their taxable counterparts.

A Word of Caution: Municipal bonds rarely are comparable to U.S. Treasury issues or government-insured bank accounts as far as credit worthiness, although these investments are frequently used as comparisons by tax-free investment brokers.

Tax Equivalent Yield

	15% State/Federal Tax Bracket	28% State/Federal Tax Bracket	31% State/Federal Tax Bracket	37% State/Federal Tax Bracket	42% State/Federal Tax Bracket
3% tax-free yield	3.53%	4.17%	4.35%	4.76%	5.17%
4% tax-free yield	4.71%	5.56%	5.80%	6.35%	6.90%
5% tax-free yield	5.88%	6.94%	7.25%	7.94%	8.62%
6% tax-free yield	7.06%	8.33%	8.70%	9.52%	10.34%
7% tax-free yield	8.24%	9.72%	10.14%	11.11%	12.07%
8% tax-free yield	9.41%	11.11%	11.59%	12.70%	13.79%
9% tax-free yield	10.59%	12.50%	13.04%	14.29%	15.52%

FIGURE 99 Here is how tax-free yields convert to taxable equivalent yields. Remember, to make this comparison valid, one must compare debts of equal credit quality and maturity. To use this table, just look for your tax bracket and then your tax-free yield. Where it crosses in the chart is what that tax-free yield is worth on a taxable bond.

ONLINE INTERNET RESOURCES

Government Statistical Data on Internet

Department of Labor–Bureau of Labor Statistics	www.bls.gov
Government Statistics in General	www.fedstats.gov/
White House Economic Briefing	www.whitehouse.gov/fsbr/esbr.html

Investment Information

SEC Edgar Corporate Filings Database	www.sec.gov/
Morningstar Mutual Funds	www.morningstar.com
Bridge Information Systems	www.bridge.com/front/
American Association for Individual Investors	www.aaii.com/
American Stock Exchange	www.amex.com/
Bloomberg	www.bloomberg.com/welcome.html
Microsoft Money Central	investor.msn.com/
Investor's Business Daily	www.investors.com/
The Wall Street Journal	www.wsj.com/
The Motley Fool	www.motleyfool.com/
Barron's Magazine	www.barrons.com/
Zacks Investment Research	www1.zacks.com/
Bank Rate Monitor	www.bankrate.com
Mutual Funds Interactive	www.fundsinteractive.com/
CBS MarketWatch	cbs.marketwatch.com
S&P Personal Wealth	www.personalwealth.com
NASDAQ Regulation: Check Your Broker	www.nasdr.com
Yahoo	finance.yahoo.com

Glossaries

Real Estate and Title Insurance Industry.	www.homeowners.com/dictionary.html
Bond Market	www.bondsonline.com/docs/ bondprofessor-glossary.html
Currency Trading	www.fx4business.com/homepage/qfgl.html
Stock Market	www.investorama.com/glossary
Business Basics	www.moneywords.com
Investments	www.investorwords.com